About Island Press

Since 1984, the nonprofit Island Press has been stimulating, shaping, and communicating the ideas that are essential for solving environmental problems worldwide. With more than 800 titles in print and some 40 new releases each year, we are the nation's leading publisher on environmental issues. We identify innovative thinkers and emerging trends in the environmental field. We work with world-renowned experts and authors to develop cross-disciplinary solutions to environmental challenges.

Island Press designs and implements coordinated book publication campaigns in order to communicate our critical messages in print, in person, and online using the latest technologies, programs, and the media. Our goal: to reach targeted audiences—scientists, policymakers, environmental advocates, the media, and concerned citizens—who can and will take action to protect the plants and animals that enrich our world, the ecosystems we need to survive, the water we drink, and the air we breathe.

Island Press gratefully acknowledges the support of its work by the Agua Fund, Inc., The Margaret A. Cargill Foundation, Betsy and Jesse Fink Foundation, The William and Flora Hewlett Foundation, The Kresge Foundation, The Forrest and Frances Lattner Foundation, The Andrew W. Mellon Foundation, The Curtis and Edith Munson Foundation, The Overbrook Foundation, The David and Lucile Packard Foundation, The Summit Foundation, Trust for Architectural Easements, The Winslow Foundation, and other generous donors.

The opinions expressed in this book are those of the author(s) and do not necessarily reflect the views of our donors.

LAWYERS, SWAMPS, AND MONEY

Lawyers, Swamps, and Money

U.S. WETLAND LAW, POLICY, AND POLITICS

Royal C. Gardner

Washington | Covelo | London

Library of Congress Cataloging-in-Publication Data

Gardner, Royal C.
Lawyers, swamps, and money : U.S. wetland law, policy, and politics / Royal C. Gardner.
p. cm.
Includes bibliographical references and index.
ISBN-13: 978-1-59726-814-1 (cloth : alk. paper)
ISBN-10: 1-59726-814-3 (cloth : alk. paper)
ISBN-13: 978-1-59726-815-8 (pbk. : alk. paper)
ISBN-10: 1-59726-815-1 (pbk. : alk. paper) 1. Wetlands—Law and legislation—
United States. 2. Wetland conservation—Law and legislation—United States. 3. Wetland mitigation banking—United States. I. Title.
KF5624.G37 2011
333.91'80973—dc22
2010041634

Printed on recycled, acid-free paper

Manufactured in the United States of America
10 9 8 7 6 5 4 3 2 1

Keywords: administrative law; Clean Water Act; wetland mitigation banking; in-lieu fee mitigation; dredge and fill; regulatory takings; Army Corps of Engineers; Rapanos v. United States; no net loss; watershed

For Mom and Dad
with love

CONTENTS

ACKNOWLEDGMENTS

Writing a book can be like wading through a swamp. It is full of surprises and discoveries, and one should not do it alone. That has certainly been the case for me.

I have been fortunate to have had a host of outstanding research assistants and Biodiversity Fellows who have helped me over the years. I especially would like to thank Marcela Bonells, Stephanie Broad, Michael Dema, Leah Ellington, Joshua Holmes, Melody James, Kristine Jones, Christine Krohn, Ezequiel Lugo, Noelle Nasif, and Michelle Sabin for their able research and dedication. Stephanie, Michael, and Marcela were particularly instrumental in finishing up this project.

I very much appreciate the support provided by Stetson University College of Law. It is a great place to work in part because of my wonderful colleagues. Dean Darby Dickerson, Ellen Podgor (past Associate Dean of Faculty Development), and Jamie Fox (current Associate Dean of Faculty Development) have always offered welcome encouragement. And the Office of Faculty Support and Stetson's law librarians have no peer.

The University of Granada Faculty of Law kindly afforded me a refuge to write during my sabbatical. My family and I treasure our time there and our dear friends, especially Antonio Sánchez Aranda and Yolanda Quesada Morillas.

Many, many people have educated me about wetland science and policy, including truly dedicated state and federal agency employees. In particular, I have benefitted immensely from working with my colleagues on the National Research Council Committee on Mitigating Wetland Losses, my previous co-authors Kim Diana Connolly and Theresa Pulley Radwan, Melanie Riedinger-Whitmore of the University of South Florida (with whom I team-teach an interdisciplinary wetland seminar), and my friends on the U.S. National Ramsar Committee and the Ramsar Scientific and Technical Review Panel. Earl Stockdale was my early mentor, and Michael Davis and Palmer Hough have always been willing to discuss and debate wetland issues with me.

Jessica Wilkinson and Michael Davis provided valuable comments on portions of drafts of this book, for which I am very appreciative.

I would also like to thank my editor, Emily Davis, and the staff of Island Press for their thoughtful guidance and advice. Any errors or omissions are of course my responsibility alone.

Finally, I would like to express my love and gratitude to my wife, Mary Fahy, and our children, Colin and Meggie. The book was a team effort. Meggie contributed mythological references, Colin suggested opening each chapter with a quotation, and Mary reviewed and provided insightful comments on innumerable drafts. I am most thankful for their love and support (and patience) during this journey.

Introduction

Wetlands pay the bills. At least that is what I have trained my two children to say. When prompted with the question, "What do we say about wetlands?" they will dutifully recite "Wetlands pay the bills," although in recent years the response is sometimes accompanied by the rolling of eyes.

I came to the world of wetlands through an unusual route. After graduating from law school, I worked in the Army General Counsel's office in the Pentagon, where I dealt with legal issues that one does not ordinarily associate with the military, such as the protection of the northern spotted owl and the administration of the Panama Canal Commission. I also served as the Department of the Army's principal wetlands attorney. While the connection between swamps, bogs, and marshes and the Pentagon may not be readily apparent, the key is the U.S. Army Corps of Engineers. Nominally a military agency, the Corps of Engineers regulates activities that affect wetlands—even activities on private property. I advised the Assistant Secretary of the Army who had the task of attempting to oversee the Corps of Engineers. (Much more on that later.) After my odd military stint, I joined the faculty at Stetson University College of Law in Gulfport, Florida, where I am fortunate to have a job that allows me to teach, research, and write about wetlands—specifically wetland law and policy. So wetlands do in fact pay my bills.

More important, wetlands also pay society's bills. Wetlands provide a host of ecosystem services, functions that benefit people. Long viewed as

1

mosquito-breeding nuisances that must be drained, wetlands have recently had their reputations rehabilitated. We now recognize that wetlands provide important habitat for animals and plants, support the seafood industry, protect homes and businesses from floods, and help improve water quality. Sadly, we often appreciate the value of wetlands and their ecosystem services only after they are gone (or degraded). Chapter 1 examines the ebb and flow of the public perception of wetlands. As society began to see wetlands less as worthless bogs and more as valuable resources, the laws governing their management also evolved.

To fully understand wetland law and policy, and to appreciate the role of politics in the process, you need to have a basic understanding of administrative law, the subject of chapter 2. Administrative law is the law that is applicable to agencies, and it is agencies such as the Corps of Engineers and the U.S. Environmental Protection Agency (EPA) that take the lead in wetland protection at the federal level. Be forewarned, however: administrative law can be a deadly boring topic. But wetland law and policy offer some interesting case studies, which should mitigate the tedium.

Chapter 3 examines the definition of a wetland—as a legal matter. You might think that what is and what is not a wetland is a relatively simple matter; however, that is most certainly not the case, at least for purposes of the Clean Water Act, the federal statute that offers the most protection to wetlands and other aquatic areas. Whether or not (and to what extent) an area is classified as a wetland can have a dramatic economic impact on a property owner, and not surprisingly the question has been heavily litigated. Through a trilogy of U.S. Supreme Court cases, we will explore the struggle to define the federal government's role in regulating wetlands. The final case of the trilogy, *Rapanos v. United States*, resulted in a fractured decision that offered no majority opinion, but gave rise to many puns (e.g., "Supreme Court muddies the waters").

Even if an area is considered a wetland for purposes of the Clean Water Act, not all activities that harm wetlands are regulated. Only the "discharge of dredged or fill material" is prohibited by the Clean Water Act, and chapter 4 considers the seemingly mundane definitions of these terms. Yet the definitions play a dispositive role in whether a mining company can engage in mountaintop removals and valley fills in Appalachia or kill a lake with semiliquid, poisonous waste from a gold mine in Alaska.

Chapter 5 turns to the unique relationship between the Corps of Engineers and the EPA. Although the Clean Water Act assigns the Corps the authority to grant or deny permits for wetland-destroying activities, Congress did not entirely trust the Corps to warmly embrace its new environmental mission. Accordingly, the EPA has a significant role in the permitting process. The Corps must apply EPA's standards, and the EPA can even veto a Corps permit. Much disagreement and frivolity ensue.

The point of the Clean Water Act and other related programs is to achieve the goal of "no net loss" of wetlands. Chapter 6 reviews some of the major threats to wetlands, including agricultural operations, home development, and invasive species. The chapter then examines various attempts to reduce wetland impacts, through providing or withholding farm subsidies, requiring Clean Water Act permittees to offset their impacts through compensatory mitigation (e.g., a wetland restoration project), and trying to convince Cajuns to eat nutria, a semiaquatic rodent. Although the federal government declared that we had achieved "no net loss" on paper, the reality is much different on the ground, especially in light of compensatory mitigation failure rates. No net loss of area does not necessarily equate to no net loss of function or ecosystem services.

One way to deal with the failure of compensatory mitigation is through wetland mitigation banking. Chapter 7 looks at a growing industry in which private companies invest money in wetland restoration projects, thereby creating wetland credits that can be sold to developers who need to offset the impacts of their projects. It is an odd "market-based" approach to environmental protection where government agencies control both the supply of and demand for the market. We will examine how entrepreneurial wetland mitigation banks work in theory and in practice.

Mitigation bankers face competition from other mitigation providers, such as in-lieu fee programs, a topic we will turn to in chapter 8. In-lieu fee programs are typically run by environmental groups, land trusts, or government agencies. Instead of investing their own money up front, the in-lieu fee administrators collect money from developers and others, pooling the funds to conduct mitigation projects in the future. At least that is the idea. We will see that many Corps districts failed to exercise proper oversight of many in-lieu fee programs, perhaps giving them a pass because their intentions were pure.

Chapter 9 provides the denouement of the conflict between mitigation bankers and in-lieu fee administrators, which came to a head when Congress directed the Corps to promulgate a regulation to level the playing field among mitigation providers. The Corps and the EPA issued a regulation that applied equivalent standards to permittee-responsible mitigation, mitigation banks, and in-lieu fee programs. "Equivalent," however, does not necessarily mean equal. And a level playing field does not necessarily rule out a preference for mitigation banks.

Chapter 10 turns to the question of enforcement. The agencies have a lot of options: administrative penalties, civil penalties, and even criminal sanctions. While the Corps and the EPA will occasionally use the enforce-ment hammer, they tend to resolve unauthorized discharges of dredged or fill material through voluntary restoration and after-the-fact permits. Citi-zens can bring a lawsuit when the government exercises its discretion not to proceed with an enforcement action, but there is a gap. Citizen suits cannot enforce permit conditions, including compensatory mitigation conditions. Moreover, there is some question about the agencies' authority to take en-forcement actions against mitigation banks and in-lieu fee programs.

Chapter 11 examines the tension between wetland regulation and pri-vate property rights. If the government physically takes your property, the Fifth Amendment to the Constitution states that you are entitled to just compensation. Sometimes the government will not physically confiscate your property, but it will limit your uses of the property, which could have a significant economic impact. At what point does a regulatory program, such as the Clean Water Act section 404 program, amount to a taking re-quiring payment of just compensation? Although the Corps and the EPA rarely lose takings cases, the specter of takings claims haunts and influences the manner in which the Clean Water Act is applied.

Looking to the future, chapter 12 makes recommendations on how we might alter our laws and policies to better protect our wetland re-sources. Finally, the epilogue provides an update on many of the wetland sites mentioned in this book. Where—or more precisely, what shape—are they in now? How have they fared as a result of our wetland laws, policies, and politics?

A few opening caveats: I have been personally involved with some of the cases discussed, from helping draft regulations that were later chal-lenged (and in some cases found by judges to be arbitrary and capricious) to participating pro bono on an amicus brief in a Supreme Court case (*Ra-panos*). I know and have had the pleasure of working with many of the play-ers mentioned (some of whom are friends). Nevertheless, I have tried to present the issues in a balanced manner.

Also, as is the case with administrative law in general, wetland law and policy can be very technical; sometimes I've covered issues in a general, sim-plified way to make the material more accessible and not get overly bogged down in details (so to speak). I suspect most readers will find it sufficiently technical, but I provide selected references and suggestions for further read-ing at the end of the book. The appendix contains relevant excerpts of the Clean Water Act, regulations, and policy documents. It is difficult to have an intelligent conversation about wetland legal matters without referring to the specific language of these texts.

Chapter 1

The Ebb and Flow of Public Perceptions of Wetlands

If there is any fact which may be supposed to be known by everybody,
and therefore by courts, it is that swamps and stagnant waters are the
cause of malarial and malignant fevers, and that the police power is never
more legitimately exercised than in removing such nuisances.
—Leovy v. United States, *177 U.S. 621, 636 (1900)*

At the beginning of the twentieth century, the U.S. Supreme Court re-
flected the common view of wetlands: they were dank, dark places that
threatened public health and welfare. In the 1900 case of *Leovy v. United
States*, the Court considered the value of land in its natural condition versus
its value in a developed state, a question that has continuing resonance today. Noting that the wet area would be worth sixty times more in agricultural production (from a mere $5,000 to a grand $300,000), the Court upheld the right of Louisiana to construct dams that dried out the swampy
lands. Indeed, the Court observed that government not only had the power
to conduct these reclamation efforts, but it was its *duty* to do so. Swamps
and their ilk were nuisances to be drained, and the newly available land
could be put to beneficial, economic use.

Wetlands have long suffered from a public relations problem. In the
legend of Hercules, he must confront the many-headed Lernaean Hydra,
whose home is a swamp. Ancient Greeks also believed that limniads
(nymphs), which inhabited marshes and swamps, would occasionally

drown people. Scottish folklore warned that the airborne fluff from marsh cattails were shape-shifted witches traveling to a secret rendezvous. The miasmic mist of swamps was once thought to cause disease and illness. Sometimes the formal names of these areas evoked dread and gloom—even hell. The military surveyor credited with naming the Great Dismal Swamp in Virginia and North Carolina certainly did not consider it to be a vacation destination. Indeed, one wetland historian observed that the word *dismal* is "derived from Dismus, the name of the thief crucified with Jesus," and thus for Christians, the word "readily signified an alliance with Satan" (Vileisis, 1997). Wetlands could be good places to dispose of garbage and bodies. The brackish marshes of New Jersey's Hackensack Meadowlands are an important Atlantic flyway for migratory birds (Tiner et al., 2002), but they are perhaps better known as trash landfills and the rumored resting place of Jimmy Hoffa.

Cultural references reinforced the notion of wetlands as public nuisances. In 1943, for example, the Walt Disney Studio produced an educational animated short entitled *The Winged Scourge*, which explains how *Anopheles* mosquitoes spread malaria. The Seven Dwarfs are then enlisted to combat this threat by destroying the mosquitoes' breeding grounds. The cartoon follows the diminutive fellows exhibiting the teamwork for which they are known: Doc and Sneezy quickly cut cattails; Happy enthusiastically spreads oil on open water; Bashful diligently sprays "a thin film" of Paris Green (arsenic and copper) on bottomland hardwoods; and even Sleepy industriously ditches and drains ponds and other waters. Oddly, Snow White does not make an appearance to supervise their work.

Books have emphasized the forbidding nature of wetlands. From classics such as *The Hound of the Baskervilles* (Grimpen Mire) to comic books like *Swamp Thing*, wetlands are portrayed as dangerous places. *Lord of the Rings* devotees (who clearly have too much time on their hands) even have a Web site devoted to the various bogs and mires of Middle Earth, none of which appears easy to traverse. The image of wetlands fared equally poorly in movies, from *Labyrinth* with its bog of eternal stench (and David Bowie as the Goblin King) to *The Princess Bride* with its fire swamp (and rodents of unusual size). Even *Monty Python and the Holy Grail* lampooned the difficulty of building in a wetland, as the King of Swamp Castle explained:

> When I first came here, this was all swamp. Everyone said I was daft to build a castle on a swamp, but I built it all the same, just to show them. It sank into the swamp. So I built a second one. That sank into the swamp. So I built a third. That burned down, fell over, then sank into

the swamp. But the fourth one stayed up. And that's what you're going to get, Lad, the strongest castle in all of England.

Of course, all does not end well for the King of Swamp Castle and the wedding party.

Often an underlying theme is that wetlands can be put to a better use. The 1957 children's book *Dear Garbage Man* perfectly captures this view of wetlands. First published ten years after Marjory Stoneman Douglas's *The Everglades: River of Grass*, *Dear Garbage Man* is a heartwarming tale of a rookie sanitation man who decides that much of the discarded refuse along his route can be recycled or reused. He embarks on a crusade to give away all his garbage, and at the end of the day his truck is empty. Alas, he discovers the following morning that the trash bins are refilled with the same items, as people found them just a tad too damaged. He consoles himself by observing that the garbage can be used to fill "lots and lots of swamps" for playgrounds and schools (figure 1-1). *Dear Garbage Man* and its anti-recycling theme enjoyed a second printing in 1988.

Recently, wetlands have become less foreboding and more hip. Wetlands Preserve was the name of an activist music club in Tribeca in New York City from 1989 to 2001 (figures 1-2 and 1-3). A documentary on the club's history, *Wetlands Preserved*, received an award for best unreleased film at the 2006 High Times Stony Awards, perhaps unfortunately reinforcing the stereotype that many environmentalists are drug-addled socialists who have no respect for private property. On the more conservative side of the political spectrum (albeit facetiously), the *Colbert Report* on Comedy Central is co-produced by Spartina Productions. *Spartina* is cordgrass, a plant species found in coastal wetlands, and each show ends by reversing the food chain as a small fish swallows a large heron or egret. There is even an adult Web site with "wetlands" in its title, although this features a markedly different type of wildlife.

The public perception of wetlands has indeed undergone a remarkable transformation. In the television series *The X-Files*, Sheriff Hartwell observes that "[w]e used to have swamps, only the EPA made us take to callin' them wetlands." As Gary Larson suggests, the term "wetland" itself projects a certain respectability otherwise lacking for the mere swamp, marsh, or bog (figure 1-4).

Rather than being viewed as mosquito-breeding nuisances (or cheap land to drain and fill), wetlands are now appreciated for the many benefits that they provide. There is a World Wetlands Day (February 2), which commemorates the signing of the Ramsar Convention, an international treaty

But suddenly, a big smile brightened his face
and he began to drag the bed to the truck.
"Start the chewer-upper, boys!" he shouted.
"All this stuff will fill in lots and lots of swamps!"
The chewer-upper started, and Emily's horseshoe
jiggled up and down. The driver leaned out and yelled,
"Stan, you're a real garbage man now!"

FIGURE 1-1. The denouement of *Dear Garbage Man*. (Text copyright © 1957 by Gene Zion. Illustrations copyright © 1957 by Margaret Bloy Graham. Renewed © 1985. Used by permission of HarperCollins Publishers.)

promoting wetland conservation, and an entire American Wetlands Month (May), which celebrates the value of wetlands.

Schoolchildren today learn the litany of wetlands functions. Wetlands, they are told, provide important habitat for fauna and flora. The list of endangered and rare species that depend on healthy wetlands runs from the perhaps still-extant ivory-billed woodpecker in the Cache River National Wildlife Refuge in Arkansas (known as the Lord God bird for the exclamation one would utter upon witnessing it) to the less well known and smaller fairy shrimp in the vernal pools of California. Wetland supporters and educators typically emphasize this function with good justification. In terms of marketing, who among us (including many developers) does not love endangered species, especially charismatic megafauna? Animals—and the

FIGURE 1-2. A Wetlands Preserve T-shirt circa 1999. (Logo used by permission of Adam Weissman.)

FIGURE 1-3. A Wetlands Preserve band. (Artist: Fred Caputi, Swampadelica Art Archives)

THE FAR SIDE® By GARY LARSON

"Well, actually, Doreen, I rather resent being called a 'swamp thing.' ... I prefer the term 'wetlands-challenged-mutant.'"

FIGURE 1-4.

cuter or bigger, the better—can help sell any product or cause. This is why one of the most popular specialty license plates in Florida pictures the native panther, all the while we build new roads, which opens new areas to development, which in turn shrinks the panther's habitat.

But wetlands benefit more than just plants and animals. People often derive benefits—ecosystem services—from wetland functions too. Indeed, the ecosystem services provided by wetlands can be of great economic

value. The commercial freshwater and marine fisheries industry needs vibrant wetlands; approximately 75 percent of commercial fish and shellfish in the United States rely on estuaries and coastal wetland systems (EPA, 2010b). Striped bass, bluefish, croaker, flounder, menhaden, sea trout, and spot are just some of the more well known wetland-dependent fish. Wetlands are important in the life cycles of anadromous species such as chinook and coho salmon, as well as catadromous species such as eel. Shrimp and crabs also spend time in estuarine and tidal wetlands during their life cycles. There is a reason why the seafood industry supported the federal Coastal Wetlands Planning, Protection and Restoration Act and other efforts to restore Louisiana's marshes. The bumper sticker that cautions "No Wetlands, No Seafood" is largely accurate.

In a related vein, wetlands also can help maintain or improve water quality. Wetland plants and soils have the capacity to remove nutrients such as nitrogen and phosphorous (as well as toxics) in runoff, thus reducing algal blooms and depleted oxygen levels in water downstream. New York City, a locale not necessarily known for its wetlands (except for the nightclub), has recognized the value of protecting wetlands in a watershed context. Rather than spending more than $3 billion in new wastewater treatment plants (with some estimates as high as $8 billion), the city decided to achieve the same level of protection by investing $1.5 billion in protecting land surrounding its upstate reservoirs (Kenny, 2006). Similarly, a key component of the Comprehensive Everglades Restoration Plan (CERP) contemplates converting agricultural lands into stormwater treatment areas (i.e., wetlands) to improve the quality of water heading toward Everglades National Park. While the CERP might fail for many reasons (such as the inability to stem development and population growth in Florida and the failure to take sea-level rise into account), the stormwater treatment concept is sound.

Wetlands can also limit damages from natural events by virtue of their flood storage and storm attenuation functions. Wetlands can act like sponges; when tides rise or rivers overflow their banks, adjacent wetlands can absorb the excess water quickly and release it slowly. When wetlands are filled or lost, the remaining area (whether developed or open water) cannot offer the same level of protection. Thus, when the record and near-record rains fell in the Midwest in 1993, the water had no place to go, as many wetlands had been converted to agricultural production. The Great Midwest Flood of 1993 caused almost $20 billion in damage in nine states and was called the "most devastating flood in modern United States history" (Kolva, 1996), at least until Hurricane Katrina in 2005. The destruction along the Gulf Coast caused by Katrina, especially the inundation of New Orleans, highlighted the vulnerability of coastal populations when their protective wetland barriers have been diminished. While intact wetland systems would not have prevented the devastation of Hurricane Katrina (or the 2004 Indian Ocean tsunami for that matter), they would have militated its effects (Ramsar Convention Secretariat, 2010a).

Finally, wetlands can serve as a refuge not only for animals but for people as well. It would be difficult to overstate the recreational value of wetlands, both in terms of aesthetics and economics. Certain segments of the public love wetlands and their wildlife. In fact, wetland-dependent birds prompt millions of people each year to tromp out to swamps, marshes, playa lakes, prairie potholes, and other wetlands. The U.S. Fish and Wildlife Service (2009) estimates that these avian aficionados spend billions of dollars annually to watch these birds or to photograph them. Or to shoot them. The "hook and bullet" crowd was among the first to recognize that protecting wetlands was in its self-interest. Fewer wetlands translate directly into fewer ducks. The Duck Stamp Program—whereby a hunter pays a licensing fee for the right to shoot a certain amount of waterfowl—was one of the early federal efforts to conserve wetlands. Today, Ducks Unlimited (a hunting organization) is one of the largest supporters of wetland restoration efforts in the United States.

So there are a host of reasons to protect wetlands, whether you are a hunter or birdwatcher, coastal resident or insurance adjuster, environmental engineer or New York City budget analyst, seafood lover or waiter, or merely a nature enthusiast. But one particular challenge to protecting wetlands (and there are many) is that in the continental United States approximately 75 percent of wetlands are in private ownership. At least this is the statistic that is commonly cited, and having been repeated enough times it has achieved an air of authenticity. Regardless of the exact percentage, however, questions about private property rights influence the debate about wetland protection: How should we as a society balance an individual's private property rights with the public benefits that wetland ecosystem services provide? What legal and policy mechanisms should we use to strike a proper balance?

In exploring these questions, we must consider the intersections of law, science, and politics. The definition of wetlands—what is and what is not a wetland—might be viewed merely as a scientific matter, but it has legal and political dimensions. Similarly, the goal of "no net loss" of wetlands might seem to be amenable to a straightforward scientific accounting, but the reality is much messier. Furthermore, who gets to make decisions about whether to permit wetland-destroying activities is far from clear and can raise fundamental constitutional issues. In some ways, wetland regulation in the United States begins and ends with the Constitution. As an initial matter, does Congress have the authority under the Commerce Clause or other provisions of the Constitution to regulate activities that damage wetlands? Or is this a responsibility under our federal system of government that must be left to state and local governments? At the end of the line, what happens when a wetland permit is denied (whether the decision maker is federal or state or local)—does the Constitution require that the disappointed property owner receive just compensation?

The study of wetland policy goes well beyond constitutional issues, however. It involves the scientific (and policy) challenges associated with endangered and exotic species, practical difficulties of enforcement actions, and political calculations. One of the most fascinating and controversial developments is the rise of wetland mitigation banking. Mitigation banking is an incentive-based, or market-based, approach to protecting wetlands. It typically involves an entrepreneur who restores a wetland, thereby generating environmental "credits" that can then be sold to developers to offset their wetland impacts. Entrepreneurial mitigation bankers and their partners are an eclectic group that includes former developers who have seen the light, environmental organizations, sod farmers, and Trappist monks.

But to fully appreciate all of these issues—from the loftiest constitutional principles to a mitigation banker's actions on the ground—you need to have a basic understanding of administrative law, a topic to which we will now turn.

Chapter 2

Administrative Law: The Short Course

Elizabeth: Wait! You have to take me to shore. According to the Code of
the Order of the Brethren . . .

Barbossa: First, your return to shore was not part of our negotiations nor
our agreement so I must do nothing. And secondly, you must be a pirate for the pirate's code to apply and you're not. And thirdly, the code
is more what you'd call "guidelines" than actual rules. Welcome
aboard the Black Pearl, Miss Turner.
— Pirates of the Caribbean: The Curse of the Black Pearl *(2003)*

If wetlands suffered from a public relations problem, its counterpart in law
schools is Administrative Law. Administrative Law is not part of the first-
year law school curriculum, like Property or Contracts. No one makes
movies like *The Paper Chase* or *Legally Blonde* about an Administrative Law
course.[1] On the surface, Administrative Law seems to lack the sexiness and
political controversies associated with Constitutional Law and Criminal
Law. In many law schools, it is not even required for graduation. Yet Administrative Law is, or at least can be, a keystone course. And it requires, as
Elizabeth in *Pirates of the Caribbean* discovered to her detriment, the ability
to understand the difference between a code or statute and mere guidance:
How are rules made and when must they be followed?

Legal education, especially in the first year, is largely mired in the common law and the study of judicial decisions.[2] Law school calls on students (literally) to scrutinize a case, recite the pertinent facts, identify the holding (or core decision) of the court, and distill its reasoning, which may be applied or distinguished in future cases. To be sure, knowledge about the common law is important for an environmental lawyer, at a minimum for historical purposes; environmental law has its origins in the common law. Legal historian Daniel Coquillette (1979) has written about *William Aldred's Case*, a 1611 nuisance case over the conversion of a sweet-smelling orchard to a noxious hog farm. *William Aldred's Case* was more than a mere property law dispute; it can be seen as an early air pollution case.

And the common law is still very much relevant today for an environmental lawyer. Much of the *Exxon Valdez* litigation over the 1989 oil spill in Alaska's Prince William Sound was based on the common law principle of negligence. (When its magnitude was stripped away, it was essentially a drunken driving case. The vehicle Captain Hazelwood was driving just happened to be carrying 53 million gallons of crude oil.) The common law is even invoked, albeit unsuccessfully so far, in climate change litigation. States and others have claimed that electric utilities and automakers' contributions to climate change amount to a public nuisance (Fuhr, 2010). But the practice of environmental law today is largely administrative and regulatory in nature.

Students take my Environmental Law course assuming that it will be about birds and bunnies. I usually wait until the third week of classes (once the add-drop period has passed) to strip away the façade and spring upon them the harsh reality that Environmental Law is basically Administrative Law in disguise. Here I have waited to do so until the second chapter.

There are a number of ways to teach Administrative Law, from focusing on the separation of powers between the branches of government to the due process rights of welfare recipients. At its heart, however, Administrative Law is about agencies, specifically executive branch agencies. For our quick introduction to administrative concepts (which focuses on federal agencies, but state agencies typically follow a similar model), we will examine (1) how agencies are structured and receive their powers; (2) what agencies do—how they make, apply, and enforce rules (binding regulations and more ephemeral guidance); and (3) how one challenges or influences agency decisions (in the courts and in other forums). Understanding administrative law is a prerequisite for understanding wetland law and policy. Wetland issues, in turn, make for fascinating case studies in administrative law.

What are agencies and who made them the boss?

As any schoolchild could tell you, the president is the head of the executive branch of government. Under the U.S. Constitution, it is the president's responsibility to ensure that the laws enacted by Congress are faithfully implemented. The president, however, is but one person. Carrying out this constitutional mandate requires a multitude of people, or, in other words, agents. And that is what most federal agencies are: agents of the president.[3]

Years ago a friend of mine taught a "Regulatory I" course, an introductory class for new employees of the U.S. Army Corps of Engineers. The Corps is an agency within the Department of Defense and, as we will see, is intricately involved with wetland regulation. Among the first questions my friend would ask the class was, who is your boss? He would receive a number of replies, ranging from the District Engineer (usually a lieutenant colonel or colonel who oversees a Corps district) to the Chief of Engineers (a three-star general in Washington, D.C.), or even the U.S. Congress (not an unreasonable response in light of Congress's history of directing the Corps to implement various pork barrel projects). Then my friend would flash up on the screen a picture of the president and explain to the new employees that this person was their boss.

While the line of authority may indeed run from a field-level Corps regulator in Omaha, Nebraska, all the way up to the president of the United States (figures 2-1 and 2-2), a different question is how these agents acquired their authority. Executive branch agencies can make rules that affect the lives of millions of people. Who made the agencies (and their employees) the boss?

The answer is both Congress and the president. Most agencies, such as the Corps of Engineers, are created by statutes, which are the result of bills passed by both houses of Congress and signed into law by the president. (A notable exception is the EPA, which came into being through an executive order signed by President Nixon.) Regardless of an agency's origin, however, its authority to act must first come from Congress. Congress must assign the agency duties or responsibilities; Congress must provide the authorizing legislation for the agency to act. Just as important, Congress must appropriate the funds that will permit the agency to try to accomplish its duties (U.S. Government Accountability Office, 2008).

So Congress has two distinct ways to control agencies. An agency may not proceed beyond what Congress has authorized, and even if Congress has authorized an action, the agency may not move forward with that

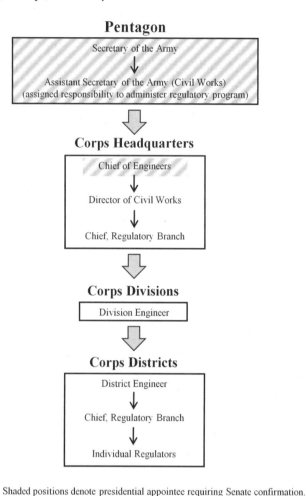

Shaded positions denote presidential appointee requiring Senate confirmation.

FIGURE 2-1. The link between the president and a Corps regulator.

action unless Congress has also agreed to pay for it. Although authorizing legislation and appropriation legislation are supposed to be separate, Congress occasionally uses the latter to direct agency actions, as we will later see.

While an agency depends on Congress for its authority and budget, the president has the greater influence, through political appointments, on how the agency will operate. Most agency employees (sometimes derisively called bureaucrats; see figure 2-3) are career or at least long-term civil servants. They bring to their jobs technical and scientific expertise; they are

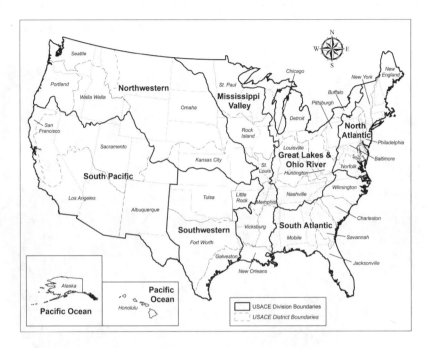

FIGURE 2-2. Corps Divisions and Districts. (Source: adapted from U.S. Army Corps of Engineers.)

not supposed to be political. But they are overseen by political appointees, people appointed by the president (and who sometimes must be confirmed by the Senate). The political appointees set the agenda and the priorities for an agency. Although Ralph Nader and others have argued that there is no difference between Republicans and Democrats, the appointments for the Secretary of Interior by Presidents Clinton and George W. Bush belie that assertion. Bruce Babbitt, the Clinton appointee, was a former governor of Arizona known for designating millions of acres of federal lands as protected areas; Gale Norton, the Bush appointee, was a former attorney for mining industry clients known for counting golf course water hazards as wetlands. The political leadership of an agency matters a great deal, to both the agency's policies and its personnel.

Political appointees may serve in Washington, D.C., or all throughout the country. An EPA organizational chart (figure 2-4) identifies the agency's top political appointees, including its regional administrators (see figure 2-5 for location of EPA regions). Note that the Assistant Administrator for Water is responsible for wetland policies.

"I'm sorry, dear, but you knew I was a bureaucrat when you married me."

FIGURE 2-3. A bureaucrat in action. (© Robert Weber/The New Yorker Collection/www.cartoonbank.com.)

Although the political appointees in charge of government agencies may change with each new presidential administration (or even midterm), the system is designed to protect the job security of career civil servants so that they are insulated from the vicissitudes of raw politics. While the rule is largely observed at the federal level, there are of course some exceptions. In the wetlands world, one of the more curious breaches occurred during Gale Norton's term as Secretary of the Interior. Michael Davis, a career civil servant for more than twenty years, worked on wetland issues in the EPA, the Corps, and the Department of Interior. As Deputy Assistant Secretary of the Army for Civil Works in the 1990s, he played a key role in designing the Comprehensive Everglades Restoration Plan and associated legislation. After enactment of the Plan in December 2000, he was the Department of Interior's lead person on Everglades restoration issues. After he apparently rankled Florida developers and farmers by pointing out the water needs of

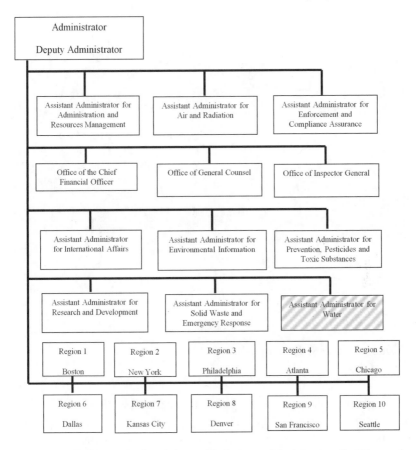

FIGURE 2-4. EPA organizational chart with Assistant Administrator for Water, who is responsible for wetland policy, shaded. (Source: EPA.)

Everglades, including Everglades National Park, he was reassigned by Secretary Norton. At the time, Secretary Norton justified the move to "reduce administrative confusion," and a 2001 *New York Times* editorial called on her to "make use of Mr. Davis's talents in an Everglades-related position in Washington." His new responsibilities, however, had nothing to do with wetlands. Indeed, Secretary Norton and her staff determined that the talents of Mr. Davis would best be used by overseeing federal desert policies. He is now with an environmental consulting firm, much happier (and politically active) in the private sector.

Although this vignette illustrates the tensions between career civil servants and political appointees, I do not mean to suggest that politics and political appointees should not play intrinsic roles in how an agency

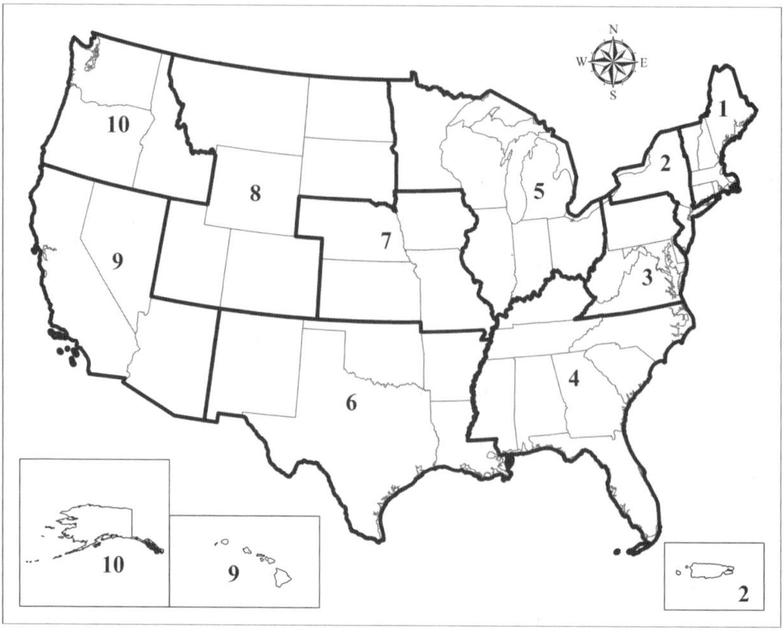

FIGURE 2-5. EPA regions. (Source: EPA.)

discharges its obligations. The regulation of private property for the benefit of the public, as well as the management of public lands for the benefit of specific individuals or groups, will always retain an element of politics. Moreover, the participation of (or at least oversight by) political appointees can provide legitimacy for an agency's actions, especially when it comes to the matter of rulemaking.

What exactly does an agency do?

Agencies make rules (an activity that is naturally called *rulemaking*) and apply them to a particular set of facts (an activity that is not so naturally referred to as *adjudication*). An agency issues rules to carry out the provisions of the statutes that it administers. The idea is that when Congress passes legislation, it is setting out broad outlines of the law—leaving it to the experts at the agency to flesh out the details. In the environmental context, an agency will apply those rules when deciding whether to grant permits to businesses and individuals to engage in some type of otherwise prohibited activity. And when the business or individual violates the rules or the permit, the agency might take enforcement actions that could result in monetary penalties (or in rare cases a prison term).

Consider, for example, the Clean Water Act, which is the federal government's primary regulatory tool to protect wetlands. In 1972, Congress enacted, over President Nixon's veto, the Clean Water Act, with its goal to "restore and maintain the chemical, physical, and biological integrity of the Nation's waters." (Despite his many faults, President Nixon was not opposed to clean water; rather, he objected to the cost of the program [Adler et al., 1993].) To achieve this lofty goal, Congress created new federal permit programs, each designed to address a different type of pollution. The National Pollutant Discharge Elimination System (NPDES) dealt with the most obvious form of water pollution, such as industrial waste spilling from the end of a pipe into a river. (The flammable Cayahoga River was still fresh in everyone's mind.) Section 402 of the Clean Water Act required that these polluters obtain permits from the EPA. Another, less obvious form of water pollution was the discharge of "dredged or fill material." Dredged or fill material can be clean dirt or sand, but it can be considered a pollutant based on its impacts. While a chemical pollutant can kill all aquatic life in a waterbody by reducing oxygen levels, fill material can accomplish the same result by eliminating the waterbody in part or entirely. To address this issue, Congress added section 404 of the Clean Water Act, which established a permit scheme for the discharge of dredged or fill material and designated the Corps as the permitting agency. (Why Congress chose the Corps for section 404 permits, and the resulting tensions that this choice created between the Corps and the EPA, will be explored in later chapters.) While both the EPA and the Corps were authorized to issue Clean Water Act permits under sections 402 and 404, respectively, the Clean Water Act itself did not precisely spell out what waters would be covered, what activities would be regulated, or what permitting standards would apply. It would be up to the agencies to promulgate rules to provide these details.

How are regulations made?

Agencies issue all types of rules, the best known of which is called a regulation. An agency regulation, when properly promulgated in accordance with the Administrative Procedure Act, has the force of law (Gardner, 1990). In other words, a regulation binds an agency and the public just as a duly enacted statute does. If you violate a regulation, you are violating the law.

Yet a regulation is not drafted by elected lawmakers; rather, regulations are generally written by career civil servants—that is, bureaucrats. While these government employees may have the scientific and technical expertise to craft an effective regulation, they are unelected. As such, they do not have by themselves the political legitimacy to issue a legally binding mandate. But Congress lacks the ability to write a statute that precisely covers every detail of implementation, and Congress therefore delegates to agencies the authority to fill in the gaps.

To resolve the concern about "regulation without representation," an agency's discretion is limited in several ways. First, political appointees oversee the work of the career civil servants. For example, the EPA Administrator and Assistant Administrators are appointed by the president. These political appointees (or lower-level appointees) will review proposed regulations, and this process provides some link between an elected official (the president, who made the appointment) and the drafter of a regulation. The Office of Management and Budget, a cabinet-level office within the White House, will also review proposed regulations, in part to ensure that they are consistent with the president's policies (Executive Order 12866, as amended, 2007).

The method by which a regulation is promulgated is another source of its political legitimacy. To produce a regulation that has the force of law, an agency almost always must engage in a public notice-and-comment process. For example, if an agency such as the Corps or the EPA wishes to issue a regulation, it must notify the public by publishing the text of the proposed regulation in the *Federal Register* (Administrative Procedure Act, 2006). The *Federal Register* is, in some ways, like the public notices published in the back of newspapers of old, only more accessible. Published daily and available on-line, the *Federal Register* contains hundreds of pages of small print announcing the availability of agency documents, the location of upcoming meetings, and the text of proposed and final regulations. When a proposed regulation appears in the *Federal Register*, various stakeholders—from regulated entities, to environmental groups, to state and local agencies—and the general public is invited to provide comments. The agency might hold a public hearing on the proposed regulation if requested, although typically it is not required to do so. After evaluating public comments, the agency then may issue the final regulation by publishing it in the *Federal Register*. In the preamble to the published regulation, the agency must respond to all substantive comments it received. Once the final regulation is published in the *Federal Register*, it becomes effective on a specified date, usually in thirty days, and is eventually codified in the Code of Federal Regulations. This public notice-and-comment process assists agencies with drafting more effective regulations, but it also helps to legitimize the role of the unelected officials who write these binding rules.

What's the difference between a regulation and mere guidance?

A notice-and-comment rulemaking process can be cumbersome and, as we will see, a rulemaking may take years to complete. Academics refer to this phenomenon as the "ossification" of rulemaking (Pierce, 1995). Rather than proceed with the notice-and-comment procedures necessary to adopt a regulation, agencies will often choose to issue "guidance" instead. An agency may give these guidance documents various titles, such as a policy statement, memorandum of agreement, regulatory guidance letter, or interpretive rule. Whatever the label, these documents share a common characteristic: they lack the force of law. Although guidance documents are rules that inform the public and interested stakeholders about how an agency will interpret and implement its governing statutes and regulations, they are not *law* in the sense of a statute or regulation. They do not appear in the Code of Federal Regulations. Accordingly, like the pirate Barbossa, agency field personnel may not always feel compelled to follow mere guidance.

There is nothing inherently improper about an agency issuing a guidance document. Indeed, Congress has authorized federal agencies to promulgate interpretive rules through the Administrative Procedure Act. Yet guidance documents do not, by themselves, create enforceable rights or obligations. Furthermore, because an agency can issue guidance documents without public notice and comment, an agency can similarly modify or revoke the rule without public notice and comment.

Many of the rules governing wetlands are found in guidance documents; these are where the real details are. The definition of the term "waters of the United States" offers a good illustration.

Navigating from statute to regulation to guidance

The Clean Water Act provides an excellent example of the relationship between statutes, regulations, and guidance. As noted above, the Clean Water Act prohibits the discharge of pollutants (including dredged and fill material) into certain waters without a permit. Specifically, the Clean Water Act prohibits discharges into "navigable waters." The statute defines "navigable waters" to be "the waters of the United States, including the territorial seas." "Territorial seas" refers to coastal waters up to three miles from shore, but what does "waters of the United States" cover? Does it include wetlands, and if so, which wetlands? Does the Clean Water Act regulate wetlands that are adjacent to what is considered a traditional navigable water (e.g., a river), wetlands that are farther upstream and adjacent to nonnavigable tributaries of these navigable waters, and wetlands that appear to have no hydrological connection to any navigable waters? Does it make a difference if the connection between the wetland and the traditionally navigable water is not through surface water, but by groundwater? Congress, at least in the plain text of the Clean Water Act, does not provide explicit answers. Thus, the EPA and the Corps filled in this gap in the law through a series of notice-and-comment rulemakings, which culminated in regulations.

The agencies' definition of "waters of the United States" is codified in the Code of Federal Regulations and lists a number of different types of waters, including certain wetlands. The regulations (33 C.F.R. § 328.3(b)) also define the term "wetlands" to mean

> those areas that are inundated or saturated by surface or ground water at a frequency and duration sufficient to support, and that under normal circumstances do support, a prevalence of vegetation typically adapted for life in saturated soil conditions. Wetlands generally include swamps, marshes, bogs, and similar areas.

While this definition suggests that a wetland has water, plants that are adapted to water, and soil that has been exposed to water, it does not necessary tell an individual property owner whether (or to what extent) his or her site is a wetland and thus subject to the requirements of the Clean Water Act. You could not take this definition out to the field and use it with any confidence to identify the dividing line between a wetland and an adjacent upland.

To help determine the boundaries of a wetland (and thus the geographic jurisdiction of the Clean Water Act), the Corps and the EPA developed wetland delineation manuals. The manuals are very technical documents; they describe indicators of the presence of water (e.g., algal mats, drift lines) and contain lists of hydrophytic vegetation (plants that are adapted to the presence of water) and hydric soils (soils exhibiting anaerobic conditions due to the presence of water). A regulator will use a delineation manual to draw lines to determine what property falls under, and what does not fall under, the Clean Water Act. The wetland delineation manuals are critically important documents for both agencies that implement the Clean Water Act and affected property owners. But they may not have gone through a public notice-and-comment process, and they are not regulations having the force of law. The manuals are guidance documents. As a legal matter, they are considered "interpretive rules," rules that explain how an agency will interpret a statute (such as the Clean Water Act) or a regulation (such as the definition of wetlands). Although they do not by themselves have the force of law, the delineation manuals can have a huge practical impact on whether a property is classified as a wetland, thus triggering the section 404 permit process. Figure 2-6 illustrates this movement from statute to regulation to guidance in the Clean Water Act context.

Statute Clean Water Act (enacted by Congress)	33 U.S.C. § 1344	Section 404: Corps of Engineers may issue permits for discharge of dredged or fill material into the "navigable waters"
	33 U.S.C. § 1362	Section 502: The term "navigable waters" is defined as "the waters of the United States, including the territorial seas"
	33 C.F.R. § 328.3(a)(3)	Definition of "waters of the United States" includes "wetlands . . . the use, degradation or destruction of which could affect interstate or foreign commerce"
Regulations Corps of Engineers Regulatory Program (promulgated through notice-and-comment rulemaking and codified in the Code of Federal Regulations)	33 C.F.R. § 328.3(a)(7)	Definition of "waters of the United States" includes "[w]etlands adjacent to [other] waters [of the United States]"
	33 C.F.R. § 328.3(b)	"Wetlands" are defined as "those areas that are inundated or saturated by surface or ground water at a frequency and duration sufficient to support . . . a prevalence of vegetation typically adapted for life in saturated soil conditions. Wetlands generally include swamps, marshes, bogs, and similar areas."
Guidance Corps of Engineers Regulatory Program (may not be subjected to notice-and-comment rulemaking and is not codified in the Code of Federal Regulations)	Technical Report Y-87-1, U.S. Army Corps of Engineers Waterways Experiment Station	Wetland Delineation Manual detailing indicators for wetland hydrology, hydrophytic vegetation, and hydric soils
	Regional Supplements	Regional supplements issued (or in process of being issued) for: Alaska; Arid West; Atlantic and Gulf Coast; Great Plains; Western Mountains; Mid-West; Caribbean Islands; and Northcentral and Northeast

FIGURE 2-6. Determining a wetland for purposes of the Clean Water Act.

What is and what is not a wetland for purposes of the Clean Water Act is the source of controversy. As we will see in chapter 3, the movement from statute to regulation to guidance to lines on the ground (or in water) is a much-litigated topic.

I'm mad as hell and not going to take it anymore: How to challenge agency actions

When an agency makes a controversial decision, the quintessential American reaction is to threaten a lawsuit. But there are procedural hurdles to filing a lawsuit (such as standing and ripeness doctrines), lawsuits are expensive and time-consuming, and at the end of the day there is no guarantee of success, as courts are very often deferential to agency decisions. Yet there are alternatives to going to court. Rather than focusing on the judiciary, it is sometimes more fruitful for citizens to seek a redress of grievances (one of the commonly overlooked rights contained in the First Amendment) through the other branches: the executive branch (and its elected officials or political appointees), the legislative branch (Congress), or even the "fourth branch," as the media likes to refer to itself. And, of course, one can always resort to litigation to seek satisfaction.

Executive Branch

Although Vice President Dick Cheney was heavily criticized for refusing to disclose which energy industry representatives he met with as part of the Bush Administration Energy Task Force (as was First Lady—later Secretary of State—Hillary Clinton pilloried when she refused to disclose who her Health Care Task Force met with), there is nothing intrinsically improper with executive branch officials meeting with industry representatives or other parties interested in the development of government regulations and policies. To be sure, sometimes such meetings might be characterized as unseemly because of a lack of transparency or the appearance that campaign contributions have purchased access or even particular policies. Yet, in a representative democracy, governmental officials should meet with their constituents.

In the wetland regulation world, one controversial example was Vice President Dan Quayle's Competitiveness Council (Percival, 2001). The council (which was essentially composed of Quayle's staff members) would review proposed agency regulations for their economic impact on business.

Part of the review would include meeting with members of the regulated community. Environmental groups (as well as some agency personnel) grumbled when the Competitiveness Council met with development industry representatives to discuss proposed wetland rules, claiming it was a back-door approach to influencing government policy. Putting aside the issue of transparency, however, there was nothing unlawful about these meetings. Indeed, such meetings can lead to more informed policy decisions. Of course, in an ideal world, an administration would seek input from all constituent groups (industry and environmental) when developing policies and regulations. But if a group feels excluded or ignored from executive branch deliberations, it can complain to Congress.

Legislative Branch

As noted above, an agency and its political appointees depend on Congress for their authority and budget. If Congress is not happy with an agency's actions, it has a number of tools to make agencies aware of its displeasure. A congressional committee can conduct an oversight hearing and call agency personnel to testify before it. Or a member of Congress can ask the Government Accountability Office (GAO) to investigate agency actions and issue a report. GAO reports can be less than complimentary, as illustrated by titles such as "Waters and Wetlands: Corps of Engineers Needs to Better Support Its Decisions for Not Asserting Jurisdiction" and "Wetlands Protection: Corps of Engineers Does Not Have an Effective Oversight Approach to Ensure that Compensatory Mitigation Is Occurring."

Congress's big hammer remains the appropriations process. Nothing gets an agency's attention like constricting its budget. Consider, for example, the House of Representatives' attempt to "zero out" the EPA's entire enforcement budget in 1995 during the Clinton administration. (To "zero out" means to allocate no money for a particular budget line item.) While this attack on the enforcement budget was ultimately unsuccessful, it partially achieved its desired effect. The EPA largely backed off enforcement for much of the 1990s (Worth, 1999).

The Media

To influence government, it is sometimes necessary to go outside of the government. Agencies pay attention to what appears in the media (both in mainstream journalism and increasingly in the blogosphere). After a 2001

U.S. Supreme Court decision called into question the extent to which federal agencies could regulate wetlands (discussed in more detail in the next chapter), the Corps and the EPA announced their intent to conduct a rule-making to clarify the geographic scope of the Clean Water Act. They solicited suggestions and comments from the public, and while the agencies were evaluating their options, one approach was leaked to the press. Purportedly, the Department of Justice was advocating an option that would have dramatically reduced the Clean Water Act's coverage. The resulting media firestorm included editorials that called on the Bush administration to reject this approach, which it did, emphasizing its commitment to the goal of no net loss of wetlands (Barringer, 2003). Whether officials within the Bush administration were seriously contemplating this option is not certain, but the public disclosure in the press forced them to clarify their position.

Judicial Branch

It is said that everyone is entitled to his or her day in court. This does not mean, however, the judge will necessarily consider every case on its merits. There are a number of procedural barriers that a plaintiff must first overcome, especially when challenging an agency decision. Chief among these are the requirements of standing and ripeness.

Standing is a complicated subject, but the general idea is that you cannot bring a lawsuit about something that is none of your business. The doctrine of standing prevents officious intermeddlers from proceeding with lawsuits against agencies, but it can also bar concerned citizens from the courts. Standing comes in two main flavors: constitutional and statutory. To bring an action in federal court against an agency or a polluter, a plaintiff must establish that it is the proper person to bring the lawsuit under both the Constitution and a particular statute.

CONSTITUTIONAL CONSIDERATIONS

According to Article III of the Constitution, federal courts may hear "cases and controversies." The U.S. Supreme Court has interpreted this phrase to mean that a federal court will only entertain actual disputes; it will not hear a complaint from a person who does not have a dog in the fight. Accordingly, the Supreme Court has stated that to establish (constitutional) standing to bring a case against an agency, a plaintiff must show three things:

1. *Injury in fact*: You need to prove that you have suffered (or are likely to suffer imminently) some specific, particularized "injury in fact." It cannot be a generalized grievance against agency policies, and it cannot be an injury suffered by the general public.
2. *Causation*: You have to establish a "fairly traceable" causal connection between the agency's conduct and the injury. The injury in fact cannot be caused by some third party who is not before the court.
3. *Redressability*: You also must demonstrate how a favorable court decision will redress the injury. In other words, if you win, will the harm be remedied? If the court's decision would have no practical or only a speculative effect, then it would amount to a mere advisory opinion.

Let's apply these concepts to a relatively simple case. Assume you are a property owner and the Corps has denied your application for a Clean Water Act section 404 permit to fill a wetland to construct a shopping mall. You have likely suffered an actual, concrete injury to your economic interests. It is a particularized injury because it affects you in a personal, individual way. The economic injury is directly traceable to the permit denial, and if a court orders the permit to be granted, the injury will be redressed. So long as you satisfy any additional statutory requirements to sue, you can proceed in court.

In contrast, it is much more difficult (but not impossible) for an environmental group to challenge the issuance of a permit to a developer. Courts are more reluctant to open their doors to allow a plaintiff to challenge an agency when the agency's alleged improper action (or inaction) is directed toward some other party. A good illustration of the difficulties faced by environmental plaintiffs is a case involving crocodiles and elephants (thereby once again invoking the charm of charismatic megafauna).

In *Lujan v. Defenders of Wildlife* (1992), environmental groups claimed that U.S. agencies had violated the Endangered Species Act (ESA) in connection with projects in Egypt and Sri Lanka. Section 7 of the Endangered Species Act requires agencies to "consult" with the U.S. Fish and Wildlife Service (FWS) to ensure that their actions do not jeopardize the continued existence of threatened and endangered species. ESA regulations (promulgated by the FWS through the notice-and-comment process) stated that consultation was not required for agency actions in foreign countries. The environmental groups focused their challenge on aid provided by the U.S. Bureau of Reclamation to Egypt for the rehabilitation of the Aswan Dam and by the U.S. Agency for International Development to Sri Lanka for another water project, which could lead, respectively, to the extinction of the Nile crocodile and the Asian elephant. The two agencies had not consulted with the FWS, and thus the specific legal question was whether the "no-consultation" regulation was consistent with the ESA. But first the environmental groups had to establish that they had standing, which they ultimately failed to do.

The U.S. Supreme Court concluded that the environmental groups had suffered no actual or imminent harm. The environmental groups had offered affidavits from their members who had previously visited Egypt and Sri Lanka and who expressed their intent to return someday. Although the Court recognized that the desire to observe an animal species is a legally cognizable interest (and that interference with that interest can establish an "injury in fact" for standing purposes), it found these "some day" intentions to return to the affected areas to be insufficient to demonstrate that the members (and thus the environmental groups) had suffered or would imminently suffer personalized harm.[4] Some members of the Court also expressed doubts about whether the redressability prong was met. If the consultation led the agencies to withdraw their support, it seemed that Egypt and Sri Lanka would move forward with the projects regardless. Thus, it was speculative that a favorable judicial decision ordering the agencies to consult with the FWS would redress the environmental groups' claimed injuries.

So the plaintiffs did not meet the constitutional requirements for standing. Where then could they turn? Although the judicial branch appears to be foreclosed as an option, they could lobby the executive branch (to convince the FWS to voluntarily conduct a rulemaking to amend the ESA regulations) and the legislative branch (to convince Congress to amend the ESA to clarify that the duty to consult applies to extraterritorial actions).

STATUTORY STANDING

Even when a plaintiff meets the constitutional requirements for standing, it also must establish statutory standing. The plaintiff generally must point to a particular statute that authorizes the lawsuit and demonstrate that it satisfies the statute's parameters. For example, the Clean Water Act authorizes private citizens to bring an enforcement action against any person "alleged to be in violation" of a section 402 permit. In *Gwaltney v. Chesapeake Bay Foundation* (1987), environmental groups sued a meatpacking plant for its discharges into Virginia's Pagan River. The meatpacking plant claimed that it had installed new pollution-control equipment and was now in compliance with its permit. The U.S. Supreme Court ruled that a lawsuit could not be brought for wholly past violations of the Clean Water Act, reasoning that the statutory phrase "to be in violation" suggested an ongoing violation. As President Clinton noted in a different context, verb tenses matter.

RIPENESS

But even if a plaintiff has constitutional and statutory standing, a court might nevertheless decline to hear a case because it is simply not the appropriate time. The claim may not yet be "ripe." As we will discuss in more detail later, in 1990 the Corps and the EPA signed a memorandum of agreement (MOA) that articulated the goal of "no net loss" of wetlands. This represented a significant shift in policy (at least for the Corps), and the MOA was issued without any public notice and comment. Industry groups immediately sued to invalidate the MOA, asserting in part that this new policy violated the rulemaking requirements of the Administrative Procedure Act. Even though the industry groups could establish that they had standing to sue, the courts still refused to consider the case on its merits. The MOA was a bit ambiguous (by design), and it was not certain to what extent Corps and EPA field personnel would be bound by the "no net loss" goal. Accordingly, in *Municipality of Anchorage v. United States* (1992), the Ninth Circuit Court of Appeals reasoned that it would be better to wait and see how the MOA was applied in the context of an actual permit application. Case dismissed.

CHEVRON DEFERENCE

Occasionally, an environmental plaintiff has standing and a court will deign to rule on the merits. A plaintiff still has an uphill battle, however, as the courts will typically defer to agency decisions. If a plaintiff challenges an agency's interpretation of a statute (contained in a regulation), the agency will likely invoke *Chevron* deference. *Chevron U.S.A. v. Natural Resources Defense Council* (1984) was an air pollution case in which the U.S. Supreme Court explained the two-part test that should be used when reviewing the validity of such agency interpretations. The first part considers whether Congress has addressed the issue in the plain language of the statute. If so, then this congressional intent must be followed. In many cases, though, the statute is not clear and congressional intent is not so obvious. The court then must consider whether the agency interpretation is a "permissible construction" of the statute. In other words, is this a reasonable interpretation of the statute? It need not be the "best" interpretation or the one that the court would necessarily choose if it were solely in charge; rather, the agency's interpretation merely must fall somewhere on the spectrum of reasonableness, taking into account the statute's language, objectives, and legislative history.

The Supreme Court has since clarified that guidance documents are not entitled to *Chevron* deference. Nevertheless, because of standing and ripeness issues, it is often difficult to challenge guidance documents until they are actually applied to a specific setting. Moreover, courts will still grant some deference to agency judgment even if it is encapsulated in a guidance document.

There are, of course, other exceptions to when *Chevron* deference is applied.[5] And, as we shall see, the Corps and the EPA's administration of the Clean Water Act section 404 program stretches *Chevron* deference to its limits, and sometimes beyond.

Chapter 3

What's a Wetland (for purposes of Clean Water Act jurisdiction)?

My position on wetlands is straightforward: All existing wetlands, no matter how small, should be preserved.

—Vice President (and presidential candidate) George H. W. Bush,
Sports Afield, *October 1988*

Following up on his campaign pledge, in January 1990 President George H. W. Bush asked his Domestic Policy Council (DPC), a group of senior advisors, to develop recommendations on implementing a policy of "no net loss" of wetlands. The DPC then embarked on a series of public meetings to solicit input and comments from various stakeholders. Obviously, a critical component to achieving the goal of "no net loss" is establishing a baseline—how many acres of wetlands remain—which naturally leads to another question: what exactly qualifies as a wetland for regulatory purposes? Did then-candidate and President Bush really mean that the federal government could and should protect all wetlands?

In May 1990, a group of federal officials visited a cornfield in Louisiana as part of an effort to evaluate how the government identified wetlands. The site had at one time been a wetland, but had been in agricultural production for decades after being ditched and drained. Yet, under a wetland delineation manual produced in 1989, it was possible to still call this area a wetland. (At a disturbed site, a regulator could rely on the presence of relic hydric soils to make this determination.) While a student or academic might marvel at the elasticity of the law that allows us to classify a cornfield as a "navigable water," it is quite a different matter to try to explain that to the farmer. Or to President Bush.

How we reached this point was an interesting administrative law journey. Several federal agencies regulate wetlands for different purposes, and over the course of years, each developed its own delineation manual or approach to classifying wetlands. The Corps used a manual issued by its Environmental Laboratory at the Waterways Experiment Station in 1987, which emphasized the three-parameter approach (hydrology, vegetation, and soils). The EPA employed its own manual, produced in 1988, which allowed greater reliance on vegetation. The FWS relied on its own classification system (Cowardin, 1979) for a national wetlands inventory, and the then-Soil Conservation Service (now Natural Resources Conservation Service) had yet another manual, which it used for wetland delineations for the Food Security Act. In a spasm of good government, the agencies decided in the late 1980s that it made common sense if all the federal agencies involved with wetlands used the same delineation manual. After a round of internal agency meetings, a joint, uniform federal wetland delineation manual was released in 1989. This manual, known as the 1989 Manual, prompted all hell to break loose.

Although wetland delineation is based on science, its application is an art. Different regulators using the same methodology can draw a wetland boundary line in different places. The 1989 Manual granted flexibility to regulators, allowing them in some cases to rely primarily on the presence of hydric soils to declare an area a wetland. The perceived effect of the 1989 Manual was that it could be used to classify more areas as wetlands, thereby dramatically enlarging the regulatory coverage of the Clean Water Act. Property owners newly subject to federal regulatory constraints were outraged—not only at the result, but by the process. And if property owners were dyspeptic, so was Congress. The agencies appeared to have expanded their jurisdiction after a series of closed meetings, and a *National Law Journal* discussed the controversy under the headline "Even the Deserts Are Wet."[1]

The adverse public reaction caught the agencies by surprise. In their view, this was a technical manual. It was an "interpretive rule," and such manuals had always been produced in this fashion. But while the agencies may have had the legal authority to issue the 1989 Manual, it was clear in hindsight that it was not a politically astute decision to proceed without involving the public.

In 1991, the Bush administration attempted to blunt the procedural objections by initiating a public notice-and-comment rulemaking to codify a new manual. A proposed rule was published in the *Federal Register*, but it too was met with a hale of criticism. The environmental community was appalled at the new methodology; it claimed that under the proposed manual, 100,000 acres of the Everglades would no longer qualify as wetlands (EDF and WWF, 1992). Developers, while generally pleased with the more restrictive approach, were nonetheless concerned about the time and expense it would take to make such delineations. The agencies received more than 100,000 public comments. In the course of considering these comments, time ran out on the Bush administration, and the Clinton administration declined to proceed with the proposal.

This particular chapter of the delineation wars was settled when, in the end, Congress reasserted its authority—through the appropriations process. In an appropriations bill, Congress specifically prohibited the Corps from spending money to implement the 1989 Manual and directed the Corps to return to the 1987 Manual (which would not cover areas such as the Louisiana cornfield). Of course, the 1987 Manual never went through public notice and comment either, but at least Congress's actions can be deemed to implicitly ratify the Corps manual.

Rather than going to Congress for relief, some property owners elected to go to court. They challenged agency decisions that their land was a wetland or that it was the type of wetland that a federal agency had the authority to regulate. Three such cases have made it all the way to the U.S. Supreme Court: *Riverside Bayview Homes v. United States*; *Solid Waste Agency of Northern Cook County v. U.S. Army Corps of Engineers*; and *Rapanos v. United States*. The trilogy illustrates different approaches to judicial review of agency actions, as well as the difficulty, as one commentator has characterized it, of "drawing lines in water" (Robertson, 2004).

The initial interpretation of "waters of the United States": We've always done it this way.

In 1972 Congress granted the Corps the authority to issue Clean Water Act permits for the discharge of dredged or fill material (which could be clean dirt or sand) into the "waters of the United States." While it may seem odd that a military agency should have such control over the use of private property, the explanation lies in history. The Corps has had a regulatory program under the Rivers and Harbors Act dating back to the 1890s. At that time, rivers were the equivalent of the nation's highways, and the Corps' job was to keep them open for military and commerce purposes. If anyone wanted to build a structure or conduct excavation activities in a navigable water, the

Corps first had to grant permission. Accordingly, when deciding which agency should be responsible for issuing "dredge and fill" permits under the new Clean Water Act, Congress took note of the Corps' decades of experience with its similar regulatory program under the Rivers and Harbors Act. (Congress, with good reason, did not entirely trust the Corps to be environmentally responsible, and thus granted some oversight authority to the U.S. EPA. The details of this awkward "power-sharing" arrangement will be fleshed out in the following chapters.)

Armed with its new authority, the Corps first needed to define its regulatory jurisdiction: what waters would be covered under the Clean Water Act? The Clean Water Act used the term "navigable waters," as did the Rivers and Harbors Act. "Navigable waters" under the Rivers and Harbors Act included only those waters that were navigable in the traditional sense: navigable in fact (used in commerce), navigable in the future with reasonable improvements, navigable in the past, and subject to the ebb and flow of the tide. So when the Corps received its Clean Water Act authority in 1972, it continued to march in the same direction, defining (through a public notice-and-comment rulemaking) "navigable waters" under the Clean Water Act as traditional navigable waters, just like under those covered by the Rivers and Harbors Act.

Not everyone agreed with the Corps' approach. Some environmental groups pointed out that although Congress used the same term ("navigable waters") in both the Rivers and Harbors Act and the Clean Water Act, Congress intended the scope of the Clean Water Act to be far broader than traditional navigable waters. As evidence, they noted that Congress specifically defined "navigable waters" in the Clean Water Act to mean the "waters of the United States" and that a congressional report expected that the term would be given "the broadest constitutional interpretation." The environmental groups sued, claiming that the Corps' restrictive regulation was not consistent with congressional intent. In 1975, in *Natural Resources Defense Council v. Callaway*, a U.S. District Court invalidated the Corps' regulation and ordered it to try again. After a series of rulemakings, the Corps (with EPA's cajoling) eventually promulgated a regulation that defined "waters of the United States" to include areas such as wetlands adjacent to traditional navigable waters and their tributaries, as well as all other isolated wetlands (having no hydrological connection to other waterbodies) that had some nexus to interstate commerce.

Specifically, in 33 C.F.R. § 328.3(a), the agency stated that

The term "waters of the United States" means

1. All waters which are currently used, or were used in the past, or may be susceptible to use in interstate or foreign commerce, including all waters which are subject to the ebb and flow of the tide;

2. All interstate waters including interstate wetlands;

3. All other waters such as intrastate lakes, rivers, streams (including intermittent streams), mudflats, sandflats, wetlands, sloughs, prairie potholes, wet meadows, playa lakes, or natural ponds, the use, degradation or destruction of which could affect interstate or foreign commerce including any such waters:

    i.   Which are or could be used by interstate or foreign travelers for recreational or other purposes; or

    ii. From which fish or shellfish are or could be taken and sold in interstate or foreign commerce; or

    iii. Which are used or could be used for industrial purpose by industries in interstate commerce;

4. All impoundments of waters otherwise defined as waters of the United States under the definition;

5. Tributaries of waters identified in paragraphs (a)(1) through (4) of this section;

6. The territorial seas;

7. Wetlands adjacent to waters (other than waters that are themselves wetlands) identified in paragraphs (a)(1) through (6) of this section.

Figure 3-1 illustrates how the regulation might play out on the landscape.

While environmentalists may have been pleased, property owners and developers were not, which set the stage for challenges to the new regulation.

United States v. Riverside Bayview Homes: Unanimity on adjacent wetlands

In 1976, Riverside Bayview Homes began preparations to construct a housing development on its property in Macomb County, Michigan, near Lake St. Clair, a traditional navigable water (see figure 3-2). Unfortunately for the company, the site contained eighty acres of marsh. Riverside Bayview Homes did not seek a Clean Water Act permit from the Corps, which obtained a court injunction ordering the developer to stop its filling of the

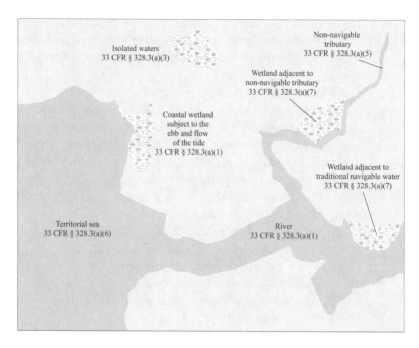

FIGURE 3-1. Waters of the United States.

marshes. The case bounced around in the judicial system for several years until in 1981 the U.S. District Court held that the property was a wetland adjacent to a navigable water, thus covered by the Corps' regulation and the Clean Water Act. The housing development could not proceed.

Undaunted, Riverside Bayview Homes appealed to the Sixth Circuit Court of Appeals, where it found a more sympathetic bench. The Court of Appeals expressed concern about the impact of the Corps' regulation on private property rights. If Riverside Bayview Homes could not construct homes on its property, the government's action could amount to an unconstitutional taking of private property; in the Court of Appeals' view, the regulation would have the same effect as if the government had physically seized the property. To avoid this potential constitutional conflict, the Court of Appeals interpreted the regulation to require frequent overland flooding from Lake St. Clair and its tributary, Black Creek. In other words, for the Corps to classify the marsh as an adjacent wetland, the area needed to have a regular hydrological surface connection with a traditional navigable water. Because there seemed to be no such connection in this case, the court ruled that the regulation (and thus the Clean Water Act) did not

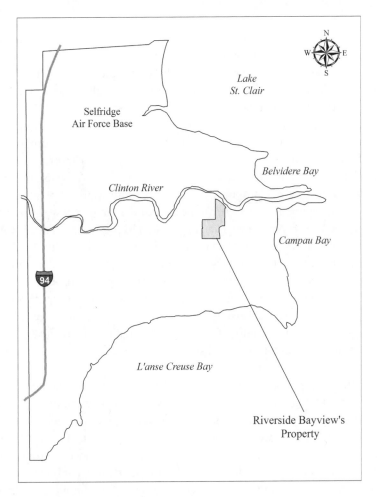

FIGURE 3-2. Location of Riverside Bayview Homes' site.

cover the marsh. The problem with the court's reasoning was that the Corps' regulation said no such thing. It did not contain a requirement of frequent inundation, and thus the Court of Appeals had essentially substituted its judgment for that of the agency.

Now it was the federal government's turn to seek further judicial review, this time before the U.S. Supreme Court. The oral argument revealed some confusion about whether the marsh had a direct surface connection to Lake St. Clair and Black Creek or whether groundwater supplied the only nexus. Indeed, the attorney for the United States admitted that "this was one of the first Section 404 cases ever tried. I don't think . . . at the time of the trial, anybody knew exactly what they were doing." With a similar surprising candor, she also conceded that "for several years the exhibits in the case were lost" and that she had not fully understood the case until they were found. Nevertheless, the Supreme Court had little difficulty reaching a unanimous decision in favor of the federal government and the Corps.

First, the Supreme Court quickly disposed of the unconstitutional takings issue that the Court of Appeals had used to justify its reading of the regulation. The mere requirement that a property owner must apply for a permit before embarking on a development project cannot, by itself, constitute a taking of private property. Why not? Because, as the Supreme Court noted (and experience has shown), the government may very well grant the permit and the property owner can proceed on its merry way. If the government denies a permit, then a takings claim may be *ripe*. But Riverside Bayview Homes had never sought a Clean Water Act permit, and thus it was premature to consider its assertion that the application of the Clean Water Act to its site significantly interfered with its private property rights. Moreover, as will be discussed in chapter 11, the proper remedy for a taking is the payment of just compensation, not invalidation of a regulation. The government can almost always take your property; it just has to pay you for it.

With the specter of unconstitutional takings and the requirement of frequent flooding removed, the Supreme Court then examined the factual question of whether the 80 acres were "adjacent wetlands" under the regulation. It was, in the Court's opinion, an "easy" question. The regulations defined wetlands as areas "inundated or saturated by surface or ground water at a frequency and duration sufficient to support, and that under normal circumstances do support, a prevalence of vegetation typically adapted for life in saturated soil conditions." The District Court had clearly found that the site contained vegetation produced by groundwater saturation; it was irrelevant that frequent flooding did not cause this hydrophytic vegetation. Furthermore, the Supreme Court observed that the wetland conditions extended to Black Creek, which itself was a traditional navigable water. Accordingly, the marsh was an "adjacent wetland."

Still, the question remained whether the regulation itself was a reasonable interpretation of the Clean Water Act. Here the Supreme Court invoked a formulation of the familiar (and deferential) *Chevron* test: "An agency's construction of a statute it is charged with enforcing is entitled to deference if it is reasonable and not in conflict with the expressed intent of Congress." Assuming that congressional intent was not clearly expressed, the Court devoted its analysis to whether it was reasonable to classify "navigable waters" to include a wetland adjacent to, but not regularly flooded by, a traditional navigable water.

While acknowledging that it might be strange to define "waters" to include land, the Supreme Court noted the difficulty of making a clear demarcation:

> The Corps must necessarily choose some point at which water ends and land begins. Our common experience tells us that this is often no easy task: the transition from water to solid ground is not necessarily or even typically an abrupt one. Rather, between open waters and dry land may lie shallows, marshes, mudflats, swamps, bogs—in short, a huge array of areas that are not wholly aquatic but nevertheless fall far short of being dry land. Where on this continuum to find the limit of "waters" is far from obvious.

Faced with such uncertainty, the Court suggested that legislative history (e.g., congressional committee or conference reports) and the goals of the Clean Water Act could provide guidance as to what is reasonable.

Improving water quality is a primary focus of the Clean Water Act, and Congress wanted to limit pollutants at their source. Recognizing that adjacent wetlands are part of a larger aquatic ecosystem, the Court observed they "as a general matter play a key role in protecting and enhancing water quality" of traditional navigable waters. An adjacent wetland may perform this water quality function regardless of whether it is flooded by a traditional navigable water. Indeed, the Court stated that "wetlands that are not flooded by adjacent waters may still tend to drain into those waters [and i]n such circumstances, the Corps has concluded that wetlands may serve to filter and purify water draining into adjacent bodies of water." Moreover, the Court recalled the important flood control and habitat functions of adjacent wetlands. Thus, it was reasonable for the Corps to define "navigable waters" to include adjacent wetlands, and the Court unanimously deferred to the Corps' "technical expertise" and "ecological judgment." Note that the *Riverside Bayview Homes* Court's paean to wetlands is a far cry from the *Leovy*'s Court denunciation of wetlands as mere public nuisances.

In a footnote, however, the *Riverside Bayview Homes* Court emphasized that this opinion applied only to adjacent wetlands. It specifically left unresolved the question of whether the Clean Water Act covered isolated wetlands, areas that had no hydrologic connection to other waters. In a bit of foreshadowing, the issue of isolated wetlands came up during oral argument. When asked about the constitutional authority for the Corps to regulate isolated wetlands, the attorney for the United States replied that the Corps' jurisdiction was based on the Commerce Clause. She said that the presence of migratory birds at an isolated wetland could provide a sufficient connection to interstate commerce. The transcript of the oral argument reports parenthetically that there was "[g]eneral laughter" (twice) in response to the concept that birds created a nexus to interstate commerce.

Solid Waste Agency of Northern Cook County v. U.S. Army Corps of Engineers: A split decision on "isolated" waters

Shortly after its victory in *Riverside Bayview Homes*, the Corps revised and reorganized its regulations, which were published in the *Federal Register*. In doing so, the Corps did not modify its definition of "navigable waters" for purposes of the Clean Water Act (that is, "waters of the United States"). The Corps did, however, break out and give "waters of the United States" its own distinct part in the regulations (Part 328). In addition, in the preamble to the regulation, the Corps attempted to clarify what it considered jurisdictional under section 328.3(a)(3): the "isolated waters," which would be regulated if their use, degradation, or destruction could affect interstate commerce. The Corps announced that it would regulate activities in isolated waters (including wetlands) that "are or would be used as habitat by other migratory birds which cross state lines." This became known as the Migratory Bird Rule (or, more derisively, as the "Reasonable Bird Test"—would a reasonable bird flying over an isolated wetland consider it suitable habitat?). The legitimacy of the Migratory Bird Rule, which had been the object of laughter during oral arguments in *Riverside Bayview Homes*, became the primary issue in *Solid Waste Agency of Northern Cook County v. U.S. Army Corps of Engineers*.

The Solid Waste Agency of Northern Cook County (SWANCC), a consortium of twenty-three Chicago-area cities and villages, banded together in the 1980s to manage garbage on a regional basis. One of its more pressing needs was to find a suitable site for a nonhazardous waste landfill. The site selected was a 533-acre parcel where sand and gravel mining operations had taken place up until about 1960. Over time, the abandoned sand and gravel pits became more than 200 permanent and seasonal ponds. Unfortunately for SWANCC, 121 species of birds found the ponds wonderful habitat, and the site was host to the second largest blue heron rookery in northeastern Illinois.

After more than six years of efforts, SWANCC had obtained all the necessary permits and approvals from the county, the Illinois EPA, and the Illinois Department of Conservation. The Illinois EPA had issued a water quality certification for the project. The last remaining hurdle was the U.S. Army Corps of Engineers. Although the ponds were not technically wetlands, they were covered by the same regulation ("other waters . . . the use, degradation, or destruction of which could affect interstate . . . commerce"), and the Corps asserted jurisdiction on the basis of the Migratory Bird Rule. In 1994, the Corps denied the permit, and SWANCC filed a lawsuit challenging the legitimacy of the Migratory Bird Rule on statutory and constitutional grounds.

The statutory argument focused on whether the Migratory Bird Test was a permissible construction of the Clean Water Act. The lower courts applied the deferential *Chevron* test. The question came down once again to one of reasonableness: was the Migratory Bird Rule a reasonable interpretation of the Clean Water Act? The Seventh Circuit rejected SWANCC's contention that the Migratory Bird Rule was an unreasonable interpretation because its focus was on wildlife, not water quality, by noting that protecting the *biological* integrity of the nation's waters was a main purpose of the Clean Water Act. Furthermore, the Seventh Circuit considered it "well established that the geographical scope of the [Clean Water] Act reaches as many waters as the Commerce Clause allows." Thus, in the court's view, the reasonableness of the Migratory Bird Rule depended on whether the presence of migratory birds established a sufficient commerce connection.

Of course, birds themselves do not engage in commerce. Yet human activities related to birds are big business. The Seventh Circuit cited Census Bureau statistics reporting that more than 3 million Americans annually spent $1.3 billion to hunt migratory birds, with at least 300,000 of those hunters crossing state lines in pursuit of their pastime and prey. Even more people (14.3 million) crossed state lines for the peaceable purpose of observing birds. Accordingly, the Seventh Circuit quickly concluded: "There is no need to dally on this point: we find (once again) that the destruction of migratory bird habitat and the attendant decrease in the populations of these birds 'substantially affects' interstate commerce."

Hunters and birders notwithstanding, SWANCC still needed someplace to put its trash. SWANCC's final judicial option was the U.S. Supreme Court. The Supreme Court agreed to hear the case, and in a 5-4 decision, reversed the Seventh Circuit. It was a stunning setback for the federal government's authority to regulate isolated waters, including wetlands.

The Supreme Court did not need to reach the constitutional question, grounding its decision on the statutory language of the Clean Water Act. In particular the Supreme Court returned to the term "navigable," emphasizing that while *Riverside Bayview Homes* said that the word had "limited import," this did not mean that "navigable" had no effect. As the Supreme Court explained:

> We found that Congress' concern for the protection of water quality and aquatic ecosystems indicated its intent to regulate wetlands "inseparably bound up with the 'waters' of the United States." . . . It was the *significant nexus* between the wetlands and "navigable waters" that informed our reading of the CWA in *Riverside Bayview Homes*. [emphasis added]

While adjacent wetlands may have a "significant nexus" to navigable waters, isolated waters by definition do not. Thus, the Migratory Bird Rule went beyond what was authorized in the Clean Water Act. It appeared that the Migratory Bird Rule had failed the first part of the *Chevron* test: when Congress has addressed the precise question, its intent must be followed.

Even if Congress's use of the term "navigable" was not clear and some ambiguity remained (as the dissent suggested), a majority of the Supreme Court was still unwilling to grant *Chevron* deference. Traditionally, the Supreme Court has been wary of affording such deference when a regulatory agency such as the Corps interprets a statute to reach "the outer limits" of congressional power under the Constitution. This is especially true, as was the case here, where "the administrative interpretation alters the federal-state framework by permitting federal encroachment upon a traditional state power," such as land-use decisions. This nondeferential approach permits the Court to make decisions based on statutory grounds and to avoid constitutional questions. With the Migratory Bird Rule dispatched by its statutory analysis, the Supreme Court did not have to reach the question of whether the federal government had the constitutional authority (under the Commerce Clause) to regulate isolated waters.

The method by which the Corps promulgated the Migratory Bird Rule may also have contributed to the Court's reluctance to grant *Chevron* deference. The Migratory Bird Rule was not a regulation; it did not, as the Court noted in passing, go through the notice-and-comment rulemaking requirements of the Administrative Procedure Act. As such, it was an interpretive rule. There is nothing intrinsically wrong with these interpretive rules, but in litigation a court may be less willing to rely on the agency's judgment if it has not been subjected to public scrutiny.

While the *SWANCC* decision might seem academically interesting for its examination of federalism, its real-world impact was stark. The Association of State Wetland Managers estimated that 40 to 60 percent of wetlands might no longer be subject to Clean Water Act protections (Kusler, 2004). While some states, such as Wisconsin, and local governments responded by enacting their own wetland legislation, other states, such as South Carolina, did not (Goldman-Carter, 2005).

From the federal government's perspective, it could have been worse. Had the Supreme Court reached the constitutional question and ruled against the Corps, the issue for all practical purposes would be closed. (It also could have had grave implications for other environmental laws such as the Endangered Species Act.) But what could the Corps and the EPA do in response in the *SWANCC* decision? They could issue guidance, but that would be granted no deference in the courts. They could proceed with a notice-and-comment rulemaking and revise their regulations, but that would take years and, even then, the courts might not be deferential. The cleanest fix would be to have Congress clarify its intent with respect to the geographic scope of the Clean Water Act, but a Republican majority in the House of Representatives rendered this option unlikely at best.

After much consideration, the Corps and the EPA labored mightily and in 2003 brought forth . . . a meek Advance Notice of Proposed Rulemaking (ANPR). An ANPR informs the public that a notice-and-comment rulemaking is contemplated, but asks the interested public for suggestions. Thus, after two years, the agencies decided only to announce that they intended to issue a proposed regulation at some point in the future, but first they needed the public's help. The agencies received thousands of comments and suggestions and ultimately abandoned the rulemaking effort. Perhaps there was no agreement within the agencies (and with the political appointees) on how to proceed. Perhaps they were embarrassed by the public reaction to the leak of one purported proposal. Perhaps the status of isolated waters was just too difficult to grapple with. Or perhaps in subsequent court cases, the agencies were winning most of the battles. It appeared that the lower courts were narrowly interpreting the *SWANCC* decision to mean that only the Migratory Bird Rule was invalid. The Corps was regulating many wetlands that were far from traditional navigable waters, but the lower courts allowed this as long as there was a hydrologic connection, no matter how attenuated it seemed.

It is important to note that isolated waters is a legal concept. Wetland scientists disagree with the premise that wetlands can somehow be isolated from the larger landscape. The law, however, likes to put things neatly into boxes or categories. Nature is a bit messier. The next case illustrates a greater divergence of law and science.

Rapanos v. United States: A trifurcation of confusion

Riverside Bayview Homes and *SWANCC* set out the extremes of the continuum of federal jurisdiction. At one end, *Riverside Bayview Homes* confirmed the federal government's jurisdiction over wetlands adjacent to traditional navigable waters, and *SWANCC* at the other end suggested that the federal government had no authority over isolated wetlands, at least under the Migratory Bird Rule. But there were a lot of waters in between. What about, for example, wetlands adjacent to nonnavigable tributaries of traditional navigable waters?

John Rapanos (with his wife and affiliated companies) owned several sites near Midland, Michigan, where he wanted to construct a shopping center. These sites contained wetlands adjacent to nonnavigable tributaries that eventually drained into traditional navigable waters 11 to 20 miles away (see figure 3-3). In 1988 the Michigan Department of Natural Resources told him that he would need a permit if he wished to fill in the wetlands as part of the project. Rapanos hired an environmental consultant, Dr. Goff, to conduct a wetland delineation of the property. Dr. Goff concluded that 48 to 58 acres were wetlands. Apparently displeased, Rapanos directed Dr. Goff to destroy the records or that he would "destroy" Dr. Goff. Rapanos then proceeded to conduct landclearing operations that destroyed 54 acres of wetlands at three sites, despite cease-and-desist orders from the Michigan Department of Natural Resources and the EPA.

Illegally filling a wetland can result in administrative, civil, and in the rarest of cases criminal penalties. Rapanos was one of these rare cases: a flagrant and intentional refusal to seek a permit for a large-scale development. The federal government brought civil and criminal charges against Rapanos, and after a thirteen-day trial, a U.S. District Court found him guilty of violating the Clean Water Act.

Rapanos's case eventually reached the U.S. Supreme Court, where once again the question was whether the Corps' interpretation of the Clean Water Act was reasonable. Unlike *Riverside Bayview Homes* (9-0) and

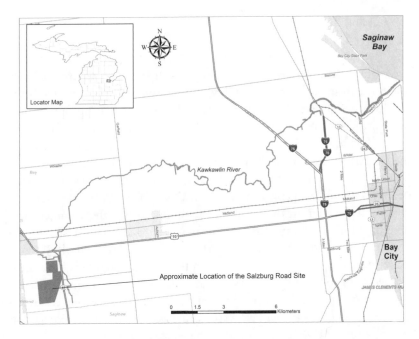

FIGURE 3-3. Location of Rapanos's Salzburg Road site. (Source: Michigan Department of Natural Resources and Environment, Land and Water Management Division.)

SWANCC (5-4), this time the justices could not agree among themselves and there was no majority opinion. Instead, the Supreme Court split 4-4-1.

Justice Scalia, writing for three other members of the Court, opened his opinion by observing that the wetland permit process was time-consuming and expensive. Citing one study, he noted that permit applicants spent more than $1.7 billion annually seeking government permission to use their property. Justice Scalia referred to the Corps as an "enlightened despot" and was hardly deferential of the Corps' position. Indeed, he was disparaging, characterizing it as "beyond parody." To underscore the risibility of the Corps' interpretation, in a footnote he quotes from the classic movie *Casablanca*:

> Captain Renault [Claude Rains]: "What in heaven's name brought you to Casablanca?"
> Rick [Humphrey Bogart]: "My health. I came to Casablanca for the waters."

Captain Renault: "The waters? What waters? We're in a desert."
Rick: "I was misinformed."

From the federal government's perspective, Justice Scalia's opinion only got worse from here. There would be no mention of a beautiful friendship.

Justice Scalia concluded that *Chevron* deference is not appropriate here because there is only one reasonable interpretation of the Clean Water Act's terms. According to Justice Scalia, "navigable waters" (and thus "waters of the United States") must refer to "only relatively permanent, standing or flowing bodies of water," such as streams, oceans, rivers, lakes, and bodies of waters that form "geographical features." In his view, this interpretation inexorably flows from the use of the definite article ("the") immediately preceding the term "waters of the United States." Had Congress simply said "water of the United States," the agencies could regulate water in general. Because, however, Congress had chosen the words "the waters," the agencies were constrained to more substantial bodies of waters. Justice Scalia's source for the plain meaning of "the waters" was his 1954 second edition of *Webster's New International Dictionary*. (One assumes that the price for this critical tool of statutory interpretation has since risen on eBay.)

With respect to whether wetlands are included within the term, Justice Scalia was just as definitive. First, to qualify as a water of the United States, a wetland must be "a relatively permanent body of water connected to traditional interstate navigable waters." Second, the wetland must have a "continuous surface connection" with the traditional navigable water, "making it difficult to determine where the 'water' ends and the 'wetland' begins." Because Rapanos's sites are far removed from a traditional navigable water, it would be unreasonable (even ludicrous) to classify them as waters of the United States.

While this restrictive and imaginative interpretation of the Clean Water Act might be reasonable (and many commentators do not think so), it is not *the* only reasonable construction. Indeed, it is not even consistent with the Supreme Court's unanimous decision in *Riverside Bayview Homes*, which specifically rejected a requirement that adjacent wetlands be connected to a traditional navigable water by surface waters. Perhaps this cramped reading of congressional intent and the dismissive view of agency expertise is why Justice Scalia was able to persuade only three of his colleagues to join him. His opinion did not carry a majority of the Court.

In contrast, Justice Stevens thought that *Rapanos* presented a "straightforward" *Chevron* question. He emphasized that the Corps has determined that wetlands adjacent to tributaries of traditionally navigable waters preserve the quality of our Nation's waters by, among other things, providing habitat for aquatic animals, keeping excessive sediment and toxic pollutants out of adjacent waters, and reducing downstream flooding by absorbing water at times of high flow.

Wetlands with even a distant hydrological connection to a traditional navigable water can provide these functions and can be properly classified as an adjacent wetland. Thus, Justice Stevens considered the Corps' construction of the Clean Water Act to be "a quintessential example of the Executive's reasonable interpretation of a statutory provision." Yet he too could convince only three of his colleagues to join him.

It fell to Justice Kennedy to break the tie. Like Goldilocks he found Justice Scalia's rigid approach to be too hard (his restrictions "are without support in the language and purposes of the [Clean Water] Act or in our cases interpreting it" and his tone is "unduly dismissive"). And he found Justice Stevens's deference too soft (his interpretation eliminates a proper tie to navigability, a "central requirement" of the Clean Water Act). But a phrase nestled in the *SWANCC* opinion—"significant nexus"—was just right.

Justice Kennedy proposed that the proper test to apply was whether the waters in question had a "significant nexus" to traditional navigable waters in light of the Clean Water Act's overall goals. The objective of the Clean Water Act is to "restore and maintain the chemical, physical, and biological integrity of the Nation's waters," and he recalled that the Corps had recognized that wetlands can play a critical role in ensuring the integrity of traditional navigable waters. He therefore proposed the following test:

> wetlands possess the requisite nexus, and thus come within the statutory phrase "navigable waters," if the *wetlands, either alone or in combination with similarly situated lands in the region, significantly affect the chemical, physical, and biological integrity of other covered waters more readily understood as "navigable."* When, in contrast, wetlands' effects on water quality are speculative or insubstantial, they fall outside the zone fairly encompassed by the statutory term "navigable waters." [emphasis added]

The only problem was that no other justice agreed with Justice Kennedy. He was the only one who embraced this "significant nexus" test.

Although the Supreme Court had splintered in its reasoning, it still had to decide what to do with Rapanos's wetlands. Justice Scalia and three members voted to vacate the lower court's judgment. Justice Stevens and three members voted to affirm. Because the lower court had not applied the "significant nexus" test to Rapanos's wetlands, Justice Kennedy joined Justice Scalia's plurality to vacate the judgment. Thus, Rapanos's conviction was overturned and the case was sent back to the lower courts for further action.

But which standard should the lower courts now apply in this case? And, more broadly, what standard should the Corps and the EPA use to assess whether other wetlands are "navigable waters" for purposes of the Clean Water Act? There was no majority opinion in *Rapanos*; the Supreme Court did not issue a single controlling opinion.

Ironically, although Justice Kennedy was the only advocate of the "significant nexus" test, it is his opinion that will be most influential. The agencies know that if a wetland meets this test and a case makes its way back to the Supreme Court, they will garner five votes upholding the federal government's jurisdiction (assuming no ideological change in the makeup of the Court): Justice Kennedy, along with four justices following the *Rapanos* dissenters' approach. As Professor Richard Lazarus observed, *Rapanos* is the *Bakke* of wetland regulation.[2]

Post-*Rapanos* response

In June 2007, the Corps and the EPA issued complicated guidance on the *Rapanos* decision. They divided "the waters" into three main categories: those over which they would definitely assert jurisdiction, those over which they definitely would not, and those that they might—if those last areas had a significant nexus to a traditional navigable water. Here are the areas the agencies considered to be "waters of the United States" under all circumstances:

- Traditional navigable waters and their adjacent wetlands;
- Nonnavigable tributaries of traditional navigable waters that are relatively permanent where the tributaries typically flow year-round or have continuous flow at least seasonally (e.g., typically three months); and
- Wetlands that directly abut these nonnavigable tributaries.

The agencies carved out a limited set of waters that they would (generally) decline jurisdiction over, which included swales and gullies with low volume or flows of short or infrequent duration, along with upland ditches with intermittent flows that drained only uplands. The significant nexus test would be applied to the waters falling between the two extremes:

- Nonnavigable tributaries that are not relatively permanent;
- Wetlands adjacent to nonnavigable tributaries that are not relatively permanent; and
- Wetlands adjacent to but that do not directly abut a relatively permanent nonnavigable tributary.

Following Justice Kennedy's lead, the agencies stated that the significant nexus analysis would focus on whether the waters in question "significantly affect the chemical, physical and biological integrity of downstream traditional navigable waters."

The federal guidance came out around the same time as the final episode of *The Sopranos*. And, in some ways, the long-awaited federal guidance issued in response to *Rapanos* mirrors the concluding scene of the HBO drama in its ambiguity. Both have been and will continue to be the subject of much debate, and both are open to varying interpretations. But while David Chase's fade to black was ambiguous by design, the federal agencies' response to *Rapanos* is ambiguous out of necessity.[3]

To say that the guidance is complex would be an understatement. It is not an easy read. In the end, it is a document that only lawyers (and law professors) can love.

But the federal agencies really had no choice; it was not possible to respond to *Rapanos* with simple, clear, crisp guidance. The bright line options were foreclosed. One bright line would have been simply to limit federal jurisdiction to traditional navigable waters (referred to in the guidance as "(a)(1) waters"). That approach, however, did not garner a majority in the Supreme Court as it was rejected by Justice Kennedy and the four dissenters. Another bright line would have been to assert jurisdiction over all waters that are hydrologically connected to traditional navigable waters. Yet that approach was rejected by the *Rapanos* plurality and Justice Kennedy. So the federal agencies were left with the task of trying to interpret the plurality's and Justice Kennedy's approaches, neither of which provided bright lines.[4] It was preordained that the guidance would be complex and confusing (and give rise to bad puns[5]).

What does the guidance mean for Clean Water Act jurisdiction? Here I must reply with the traditional law school answer: It depends. It depends on how the guidance is interpreted and applied in the field. The Corps is a decentralized agency, and individual districts (and individual regulators) may apply the "significant nexus" test broadly or narrowly. It depends on the amount of resources the agencies have, which in turn will influence how much data the agencies can collect. It depends on how aggressively the regulated community challenges assertions of jurisdiction. It also depends on whether the environmental community challenges decisions not to assert jurisdiction.

Like *The Sopranos*, which is now consigned to expurgated reruns on the A&E network, the saga of federal wetland jurisdiction will live on and on. Ideally, Congress will step in to resolve the ambiguities. Then we can move on to debate and litigate the constitutional issues that were not reached in *Rapanos*.

The constitutional limits of the Clean Water Act

What if Congress stepped into the breach and sought to claim federal jurisdiction to the farthest extent permitted by the Interstate Commerce Clause? How far upstream from traditional navigable waters could the federal agencies then regulate? The answer may lie in two U.S. Supreme Court cases dealing with two distinctly different types of crops: wheat and marijuana.

In the 1942 case of *Wickard v. Filburn*, the Supreme Court considered the constitutionality of New Deal legislation that regulated the amount of wheat a farmer could grow. Concerned about the price and supply of wheat, the federal government even restricted the amount of wheat a farmer could harvest for use on his own farm. One farmer challenged this federal intrusiveness as beyond Congress's Commerce Clause powers. The Supreme Court, however, upheld the legitimacy of the law by applying a cumulative impacts analysis. The Court reasoned that this purely local activity "may still, whatever its nature, be reached by Congress if it exerts a substantial economic effect on interstate commerce." Although the impact of a single farmer's activities would be de minimis, the Court nevertheless held that the potential cumulative impact of such trivial actions could be covered by the Interstate Commerce Clause. What if everyone grew their own wheat? Why, that could wreak havoc on wheat demand, supply, and prices.

For many years, *Wickard* was viewed as an outlier. It was studied in law schools as an example of the most extreme reach of the Congress's commerce power, but it was not frequently cited with favor by the Supreme Court. At least that was the situation until *Gonzales v. Raich*, the 2005 medical marijuana case that breathed new life into the cumulative impacts analysis.

If one puts aside the emotional issues of illegal drug usage and cancer patients, at its heart *Raich* is a Commerce Clause case. Does Congress have the authority to prohibit under the Controlled Substances Act the possession of homegrown marijuana even if it is intended for personal, medical use? The Supreme Court ruled that Congress could indeed regulate a purely intrastate activity in an attempt to limit the growth of a national, interstate market. Citing *Wickard* with approval, the Supreme Court found that precedent controlling: "leaving home-consumed marijuana outside federal control would similarly affect price and market conditions."

The connection between medical marijuana and wetlands may not be immediately obvious (especially since marijuana, shown in figure 3-4, is a facultative upland species). But the reasoning of the *Raich* Court, and its reaffirmation of the cumulative impacts analysis, could be used to justify the Migratory Bird Rule.

In *SWANCC*, the Supreme Court invalidated the Migratory Bird Rule as going beyond congressional intent. By deciding the case on statutory grounds, the Court avoided reaching the constitutional issue, thus leaving

FIGURE 3-4. A facultative upland species with implications for federal wetland regulation. (Used by permission of iStockphoto)

open the possibility that migratory birds may provide a sufficient interstate commerce nexus to assert federal jurisdiction over isolated wetlands.

The cumulative impacts analysis used in *Wickard* (what if everyone grew their own wheat?) and *Raich* (what if everyone grew their own marijuana?) is transferable to isolated waters that provide habitat for migratory birds. If a few hydrologically isolated wetlands are filled, the impact on migratory birds will not be substantial. But the cumulative impact of filling thousands of acres of these areas would have a significant impact on migratory bird populations—and a substantial impact on interstate commerce. A recent FWS economic analysis of birding in the United States demonstrates the link between wetland bird species and billions of dollars of bird-related economic activity. In 2001 birders spent nearly $32 billion in retail purchases, and 47 percent of them visited wetlands to conduct birding activities. If the wetlands go, so do the migratory birds and the commercial activities dependent on them.

This constitutional issue will not, however, be squarely presented unless Congress amends the Clean Water Act to clearly assert jurisdiction to the fullest extent permitted by the Commerce Clause. While Congress occasionally considers bills to do just that (e.g., removing the term "navigable waters" from the Clean Water Act), the legislative process is laborious and the result almost always contains compromises and ambiguities. Chapters 4 and 5, which examine the roles Congress assigned to the EPA and the Corps in the wetland permitting process, starkly illustrate this point.

Chapter 4

Dredge and Fill: The Importance of Precise Definitions

There are strange things done in the midnight sun
By the men who moil for gold . . .
—*Robert Service, "The Cremation of Sam McGee" (1907)*

The allure of gold has long caused misjudgments and madness, so perhaps it is no surprise that today's regulation of mining activities is a bit peculiar as well. The land of the midnight sun is not the only place where strange things are done. Washington, D.C., is also a prime location for oddities, and the Clean Water Act is a leading example. The statute, which is the federal government's strongest regulatory tool to protect wetlands, does not prohibit all activities that harm wetlands. It is not illegal under the Clean Water Act to remove wetland vegetation. It is not illegal under the Clean Water Act to excavate or dredge wetlands. It is even not illegal under the Clean Water Act to drain a wetland. Rather, what is prohibited is the "point source discharge" of "dredged or fill material." Thus, it is of paramount importance how these terms are defined, and the definitions have an impact on development and mining operations from Appalachia to Alaska.

A lesson for young lawyers: Read the statute.

One of the first things that a lawyer must do before advising a client is to determine the applicable law. What law governs your client's actions? Are they covered by a particular statute or statutes (and the implementing regulations)? You might assume that if your client is doing work in a wetland that meets the definition of "waters of the United States," then the Clean Water Act must be triggered. But that assumption may be incorrect. Geographic jurisdiction (what waters are covered) is only one part of the equation. The activity must also be the subject of regulation, and the Clean Water Act uses very precise language in this regard. Only "point source discharges" trigger the need for a Clean Water Act permit.[1]

The Clean Water Act defines "point source" to be "any discernible, confined and discrete conveyance." The statute provides an illustrative (but not exhaustive) list of examples, including pipes, ditches, channels, and tunnels. "Discharge" of a pollutant is defined as an addition of a pollutant from a point source. Does this definition cover the removal of wetland vegetation (landclearing), the removal of wetland soils (excavation or dredging), or the removal of water from a wetland (draining)? At first glance, the answer would seem to be no: a removal of something, whether plant, soil, or water, is not an "addition." But, as you should know by now, the answer is a lawyerly "it depends"—it depends on how one does the removing.

Does landclearing require a Clean Water Act permit?

In the late 1970s, a farmer in Louisiana wanted to grow soybeans on a 20,000-acre tract of land. The land, however, was mostly bottomland hardwood swamp. So after loggers came in to harvest the commercially valuable trees, the farmer conducted a landclearing operation to prepare the area for soybeans. He employed workers who used bulldozers and backhoes to cut the remaining timber and vegetation. The debris was then mechanically raked into windrows, which were burned. The effect of these activities was to level the land. Environmental groups, along with local fishing and hunting organizations (which had established standing by claiming that the destruction of the wetlands would adversely affect local fish and wildlife), brought a lawsuit to halt the landclearing operation without a Clean Water Act section 404 permit. While the farmer's actions were indeed destroying thousands of acres of wetlands, the key legal question was whether there were any point source discharges of dredged or fill material.

Ultimately, in 1983, in *Avoyelles Sportsmen's League v. Marsh*, the Fifth Circuit Court of Appeals concluded that a permit was needed. The blades of the bulldozers and shovels of the backhoes moved material (soil and vegetation) from place to place, and the court found these machines to be point sources—that is, discrete conveyances. But were they discharging dredged or fill material? The material, after all, originated on the property itself. What was the "addition"? Pointing to the overall goal of the Clean Water Act, the court held that term "addition" in the Clean Water Act "may reasonably be understood to include 'redeposit.'" So the last issue was whether the point sources (bulldozers and backhoes) were discharging (re-depositing) dredged or fill material: did the soil and vegetation from the wetland itself constitute *dredged or fill material*? At the time, the Corps' regulations defined "fill material" to be material whose primary purpose was to replace a water with dry land or to raise a waterbody's bottom elevation. If the primary purpose of discharging the material was simply to dispose of waste, then it was not considered to be fill material (and would be regulated by the EPA under Clean Water Act section 402). Applying the Corps' intent-based definition of fill material, the court noted that the movement of the soil and vegetation had the effect of replacing a wetland with dry land, which was the farmer's objective all along. Thus, the farmer was discharging fill material.

The Corps responded to the decision by issuing to its field staff a Regulatory Guidance Letter (abbreviated as RGL and pronounced "regal"). The RGL reiterated the court's holding—that mechanized landclearing activities in wetlands can constitute point source discharges of fill material, requiring a Clean Water Act section 404 permit—and announced that it would be applied beyond the Fifth Circuit (which covers Louisiana, Mississippi, and Texas) to the entire country. The RGL stated, however, that a 404 permit was not needed for de minimis discharges of dredged or fill material and that the mere removal of vegetation was not a regulated activity. So cutting a tree with a chain saw, even if the tree fell in a wetland, did not fall under the Clean Water Act's purview.

Does dredging (and sidecasting) require a Clean Water Act permit?

Dredging is the act of excavating or digging up soil from the bottom of a waterbody. Even though the Clean Water Act requires a permit for the discharge of dredged material, it does not technically regulate dredging itself. Only the dumping (discharging) of the dredged or excavated material into waters of the United States triggers the need for a Clean Water Act section 404 permit.

But it is hard to conduct a dredging operation without discharging the excavated material into waters. Dredging is commonly required to deepen a harbor or to keep a shipping channel open. There are several different

FIGURE 4-1. The *Merritt*, a sidecaster in the Corps' Wilmington District. (Source: U.S. Army Corps of Engineers.)

types of dredges, but all collect sediment from the water's bottom, either by a suction pipe or buckets. In some vessels, the excavated material, frequently referred to as dredged spoil, is loaded into a hopper bin and later transported to an aquatic dumping site. In shallow-draft areas, the dredge may not retain the dredged spoil, but instead discharge it through an elevated boom (see figure 4-1). The process of scooping or sucking up the sediments and then discharging them through a boom is known as sidecasting, because the material is essentially cast to the side. The sidecasted material may land elsewhere in the waterbody being dredged, or on the shore or in adjacent wetlands.

If the dredged spoil lands in water, then there is a point source discharge of a pollutant into navigable waters. The point source is the elevated boom (a discrete conveyance), the discharge is the addition (redeposition) of dredged spoil, and the navigable water is the river, channel, harbor, or adjacent wetland (and we know it's navigable in the classic sense; the commerce connection is evident from the presence of the dredge itself). And the pollutant, of course, is the dredged spoil, even if it is just clean sand, uncontaminated with toxics. The Clean Water Act defines "pollutant" to include "dredged spoil." In fact, "dredged spoil" is the first example cited in the definition. The act of removing the sediment from the water's bottom can transform it, from a legal perspective, into a pollutant.

The Corps has consistently required section 404 permits for the sidecasting of dredged spoil (a point that was reaffirmed in the *Avoyelles* RGL).

The sidecasting rule applied to developers who ditched a wetland to drain it. If the wetland soils (now transformed into dredged material) were sidecast into the wetland, the developer needed to apply for a section 404 permit. But what if the developer did not sidecast the dredged material—could the Corps still require a 404 permit?

Neatness counts: Exploiting a loophole

A developer in North Carolina wanted to develop 700 acres of pocosin wetlands, which are evergreen shrub bogs on the coastal plain (Richardson, 1983), but did not want to go through the section 404 process. So he ditched the exterior of the site, but did not sidecast the dredged material. Instead, he loaded it up on dump trucks and had it carted off to an upland disposal site. While this process was expensive, the developer calculated that it was less expensive than applying for a permit and waiting for a decision (and providing compensatory mitigation to offset the project's impacts). The Wilmington District of the Corps observed the ditching, but concluded that there were only de minimis discharges and thus no grounds to assert its regulatory authority.

The ditch had the effect of drawing water off the site's interior and killing off the wetland plants. Once the site no longer exhibited wetland hydrology, the developer asked the Corps for a jurisdictional determination. It was no longer a wetland, the Corps declared. The developer proceeded with building homes where the pocosin wetlands once resided.

Unhappy with this turn of events, the North Carolina Wildlife Federation and the National Wildlife Federation filed a lawsuit against the Corps and the EPA, naming Wilmington District Engineer Colonel Walter Tulloch as the lead defendant, for the failure to require a Clean Water Act section 404 permit. (Colonel Tulloch was sued in his official capacity; thus, he would not be found personally liable.) The funny thing here was that many EPA and Corps (and Pentagon) headquarters staff agreed with the environmental groups. It was outrageous that a developer could get around the Clean Water Act by simply carting off the dredged material. The effect was the same as if the material had been sidecast: the entire area had been drained. The agencies viewed the litigation as an opportunity to tighten up some regulatory loopholes.

So, with the approval of political appointees in the EPA and the Pentagon, the Corps and the EPA settled the lawsuit and agreed to conduct a rulemaking to revise the regulatory definition of "discharge of dredged material." The settlement agreement called on the agencies to issue a proposed rule to broaden the term to include "incidental fallback." Incidental fallback occurs when a clamshell bucket scoops up the sediment; inevitably, some of the sediment (now considered dredge material) drips off the bucket and falls back into the water. Those drippings—the incidental fallback—were the legal hook to assert Clean Water Act jurisdiction over the entire activity. The settlement agreement did not promise specific language in the final rule, however. To do so would have run counter to the rulemaking requirements of the Administrative Procedure Act. The notice-and-comment process (and the settlement agreement) required the agencies to consider and respond to substantive comments about the proposed rule from stakeholders, such as the National Association of Home Builders, in good faith. Disregarding those comments would make a sham of the process and leave the agencies vulnerable to a legal challenge on procedural grounds.

After the notice-and-comment process, the agencies did decide to regulate incidental fallback. They defined "discharge of dredged material" to include "any addition, including any redeposit, of dredged material, including excavated material, into waters of the United States which is incidental to any activity, including mechanized landclearing, ditching, channelization, or other excavation." The final rule (which would be codified in the Code of Federal Regulations) contained an exception for incidental additions that did not have the effect of degrading or destroying waters of the United States. Accordingly, as the preamble to the final rule helpfully explained, the Corps would not regulate someone "walking, bicycling, or driving a vehicle through a wetland." (Some commenters questioned whether the proposed rule was so broad that it would cover wetland soil attached to one's boots or a tire's treads.) The final rule essentially created an effects-based test for regulated activities and was known as the Tulloch rule, in honor of Colonel Tulloch. One long-time agency official characterized the Tulloch rule as the most significant environment improvement that he had ever been involved with. The mining companies and others in the regulated community were not so enthusiastic.

The inevitable blowback: The regulated community responds.

The agencies faced a second round of litigation on a different front, this time from disgruntled development interests. The case, *National Mining Association v. U.S. Army Corps of Engineers*, made its way up to the U.S. Court of Appeals for the D.C. Circuit. There in 1998 a three-judge panel unanimously rejected the Tulloch rule not as procedurally flawed, but on the substantive ground that it was an invalid interpretation of the Clean Water Act.

The court focused on the text of the Clean Water Act and its relationship to other Corps authorities. The court emphasized that the statutory definition of "discharge" was "addition." Judge Williams, writing for the court, held that it was manifestly unreasonable for the agencies to interpret the term "addition" to include "the situation in which material is removed from the waters of the United States and a small portion of it happens to fall back." He stated that "Congress could not have contemplated that the attempted removal of 100 tons of [dredged spoil] could constitute an addition simply because only 99 tons of it were actually taken away." (The agencies were so out of bounds he apparently did not feel the need to formally apply the *Chevron* test.)

The court then contrasted the Corps' section 404 authority with its authority under section 10 of the Rivers and Harbors Act of 1899. Section 404 covers the discharge (addition) of dredged or fill material, while section 10 requires a Corps permit for the excavating or filling of traditional navigable waters. Accordingly, because section 10 employs the term "excavate," the court concluded that removal activities are governed by the Rivers and Harbors Act, rather than the Clean Water Act. The thought was that Congress knows how to regulate excavation activities if it wanted to (notwithstanding the fact that there was a span of seventy-three years between the passage of the two acts). The court reasoned that any attempt to bring the two acts into harmony must be done by Congress. In the court's view, the Corps and the EPA could not "do it simply by declaring that incomplete removal constitutes addition."

In a concurring opinion, Judge Silberman noted that the term "addition" conveyed "both a temporal and geographic ambiguity." Whether a redeposition of dredged material was an "addition" depended on how long it had been removed from the water and how far it had been transported. Because of this ambiguity, he invoked the *Chevron* analysis, under which the Tulloch rule would be upheld unless it was an impermissible construction of the statute. Judge Silberman found the Tulloch rule to be an unreasonable interpretation because subjecting all dredging activities to Clean Water Act regulation was inconsistent with the structure of that law, as well as the Rivers and Harbors Act. Furthermore, he pointed out that section 404 refers to permits for discharges at "specified disposal sites," a statutory term that did not seem to contemplate incidental discharges, where the disposal site is essentially the same place from which the material originated.

The impact of the Tulloch rule's invalidation varied from Corps district to Corps district. Some districts, such as the Nashville District, rarely relied on the Tulloch rule, while others, such as the Los Angeles District, reported that as much as 44 percent of its Clean Water Act work depended on the Tulloch rule (Gardner, 1998). Several years later, the U.S. Supreme Court had the opportunity to weigh in on the issue of what constitutes an "addition."

Deep plowing or deep ripping? The *Borden Ranch* case

A landowner in California wanted to convert his property from ranchland and farmland to a vineyard. To allow for the root growth of his new crop, he needed to break up a clay layer beneath the topsoil. He accomplished this task by "deep plowing," which involved a bulldozer or tractor dragging four- to seven-foot tines that gouged through the clay. The Corps and the EPA called it "deep ripping." When the landowner began breaking up the clay underlying vernal pools, swales, and intermittent streambeds (thus draining the areas), the Corps asserted jurisdiction over the site and the activities. After several years of negotiations and after the landowner ignored Corps and EPA orders to stop ripping up the wetlands, the matter wound up in court. The agencies sought monetary penalties and restoration; the landowner claimed the federal government had no business regulating how he plowed his land and that simply churning up soil was not an "addition" of a "pollutant."

In a 2001 decision, the Ninth Circuit Court of Appeals ruled 2-1 in favor of the agencies in the case of *Borden Ranch Partnership v. U.S. Army Corps of Engineers*. The majority found that deep ripping (as the court characterized it) was analogous to sidecasting and cited *Avoyelles* for the proposition that a redeposit could be an addition. Noting that deep ripping caused soil to be "wrenched up, moved around, and redeposited somewhere else," the court remarked that "activities that destroy the ecology of a wetland are not immune from the Clean Water Act merely because they do not involve the introduction of material brought in from somewhere else." The dissenting judge thought the case was more akin to incidental fallback, which the D.C. Circuit had held in *National Mining Association* could not be regulated under the Clean Water Act. While noting that "deep plowing" (as he referred to the activity) transformed the ground and hydrological regime, he disputed that there had been any "addition." Because "Congress spoke in terms of discharge or addition of pollutants, not in terms of change of the hydrological nature of the soil," he contended that the Clean Water Act did not reach this activity.

The U.S. Supreme Court granted certiorari to hear the case, which was not a good sign for the agencies. (The Ninth Circuit Court of Appeals has the reputation for being the circuit court whose decisions are most often reversed by the U.S. Supreme Court.) During oral argument, the justices considered what would happen if the parties' arguments were carried to their logical extreme. The attorney for the landowner was asked whether punching a hole and draining Lake Erie would implicate the Clean Water Act. He replied no—with no addition of a pollutant, there would be no jurisdiction regardless of the environmental impact (although he tried to note that excavating and draining Lake Erie would be regulated under the Rivers and Harbors Act). The attorney for the U.S. government argued that moving wetland soil a few inches vertically or horizontally was sufficient to constitute an addition. He then had to parry questions about whether boots could be a conveyance (a point source) and whether raking a beach—with the attendant movement of the soil—would be an addition of a pollutant.

It looked to be a close decision, and in many 5-4 cases Justice Kennedy would supply the decisive vote (recall *SWANCC* and *Rapanos*). But in this case, Justice Kennedy was a friend of the landowner Angelo Tsakopoulos, so he recused himself from the case. Without Justice Kennedy's participation, the Supreme Court was deadlocked with four votes to affirm and four votes to reverse. Accordingly, in 2002 the Court issued a one-sentence per curiam opinion, noting that the decision below was affirmed by an equally divided court. Tsakopoulos reportedly observed that "it's a pretty expensive friendship I have with Justice Kennedy" (Bishop, 2004).

In such a circumstance, the lower court's decision remains in effect, but it does not set precedent for the entire country. Whether deep ripping or deep plowing results in an addition of a pollutant remains an open question. Given how far the agencies stretched the definition of "addition," they will likely be careful to avoid litigating the issue in the future.

Fill, baby, fill

The reason that the Corps and EPA tried to regulate such minor discharges was that the activity causing the discharge could have a dramatic and devastating environmental impact (unconnected to the discharge itself). Thus far we have largely focused on the discharge of dredged material. Let us now turn to its counterpart, the discharge of fill material.

Initially, the Corps and the EPA shared the same definition of "fill material": "any pollutant used to create fill in the traditional sense of replacing an aquatic area with dry land or of changing the bottom elevation of a water body for any purpose." The Corps became concerned that this definition could include the discharge of solid waste materials that would be more appropriately under the purview of the EPA, not the Corps and section 404. So the Corps proceeded in 1977 with a notice-and-comment rulemaking that amended the definition to adopt a "primary purpose" test. The Corps' regulation then stated that "fill material" refers to:

> Any material used for the primary purpose of replacing an aquatic area with dry land or of changing the bottom elevation of an [sic] water body. The term does not include any pollutant discharged into the water primarily to dispose of waste, as that activity is regulated under section 402 of the Clean Water Act.

In contrast, the EPA's definition considered material to be fill if it replaced waters with dry land or raised a bottom elevation for any purpose. The Corps had a purpose-based test, while the EPA had an effects-based test.

Not surprisingly, the disjointed definitions gave rise to problems in the administration of the Clean Water Act. One area where this conflict played out was in the coal mines of Appalachia. Mining companies had developed a new method of extracting coal, which became prevalent in the mid-1990s. Rather than tunneling underground, miners would set explosives to blast the top of mountains. They would then remove the rock and soil, known as overburden, to get to the low-sulfur coal, which was becoming increasingly more in demand. (Burning low-sulfur coal results in fewer carbon emissions than burning high-sulfur coal.) Mountaintop removal mining was more efficient and safer than traditional coal mining.

Once the mountaintop was blown off and the coal removed, however, there was the problem of what to do with the overburden. Once released, the previously compressed material expands and cannot simply be placed back where it was. Even with the coal removed, the mined area can no longer accommodate the overburden. The easiest technique for disposing of the overburden was dumping it in the valleys (with their streams and wetlands) below (see figure 4-2). But was the overburden "fill material" for purposes of the Clean Water Act, in which case the Corps would be the permitting agency under section 404, or was it simply a waste product of the mining operation, in which case EPA would be the permitting agency under section 402? The coal companies favored the former—in part because the Corps would be more likely to grant the permits and to do so in an expeditious fashion through a general permit known as nationwide permit (NWP) 21.

FIGURE 4-2. An example of mountaintop removal mining, southern West Virignia. (Source: Ohio Valley Environmental Coalition. Photo credit: Vivian Stockman. Flyover courtesy: SouthWings.org.)

Mountaintop removal and nationwide permit 21

The Corps issues two types of section 404 permits: individual permits and general permits. Corps districts issue individual permits after a case-by-case review of the proposed project and its environmental impacts. (The details of the individual permit process will be reviewed in chapter 5.) General permits, which can be authorized at the headquarters or regional level, authorize minor impacts to waters of the United States. General permits are a blanket form of advance authorization for a category of activities. With some general permits, the permittee does not need to notify the Corps prior to construction. As long as the project falls within the general permit's parameters (e.g., a boat ramp that is less than 20 feet wide requiring less than 50 cubic yards of fill material), the permittee can proceed with the project. Other general permits (e.g., linear transportation projects affecting a half acre of nontidal wetlands) require the permittee to contact the Corps beforehand and comply with certain conditions. In either case, Clean Water Act section 404(e) limits the use of general permits to a category of activities that "will cause only minimal adverse environmental effects" on an individual and cumulative basis.

The advantage of a general permit to a permittee is obvious. The project can move forward much more quickly and with less agency review than it would under a cumbersome individual permit review. Of particular interest here is NWP 21 for surface coal mining operations (it is called a "nationwide" permit because it is a general permit issued by headquarters and is applicable throughout the nation).[2] Hundreds of mountaintop removal operations have been authorized under NWP 21, and the loss of aquatic resources has been great. Indeed, just four West Virginia mining projects authorized under NWP 21 resulted in twenty-three valleys filled and the destruction of more than 13 miles of headwater streams, which appears to violate the plain language of section 404(e) that limits the use of general permits to activities resulting in only minimal impacts.

And environmental organizations have made such arguments in trying to end what some call the "Appalachian Apocalypse." While they had some initial success at the trial level (even obtaining an injunction prohibiting the use of NWP 21), the Fourth Circuit Court of Appeals eventually deferred to the Corps' minimal-effects determination (*Ohio Valley Environmental Coalition v. Aracoma Coal Co.*, 2009). Another ground for attack was the issue of whether overburden was truly fill material under the Corps' primary purpose test. If the overburden was characterized as waste, then the permitting decision would move from the Corps under section 404 to the EPA under section 402. The EPA would be less likely to grant the permits for mountaintop removal and valley fills, or at least it would conduct an individual review of each mining operation.

To fend off these challenges, during the Clinton administration, the Corps began a notice-and-comment rulemaking in April 2000 to modify its definition of fill material. The primary purpose test was to be replaced with the EPA's effect-based approach, and the proposal made it explicit that the "placement of coal mining overburden" was a discharge of fill material. The Bush administration completed the rulemaking in May 2002, issuing a final rule that expanded the examples of fill material to include "rock, sand, soil, clay, plastics, construction debris, wood chips, [and] *overburden from mining or other excavation activities*" (emphasis added). Thus, the permitting for valley fills from mountaintop removals will remain with the Corps and the 404 program.

The modification of the definition of fill material also benefitted mining operations elsewhere. A 2009 U.S. Supreme Court case involving a gold mine in Alaska underscores the importance of the definition of such technical terms.

Strange things done in the midnight sun: Gold mining waste as fill

A mining company, Coeur Alaska, wanted to reopen the Kensington gold mine in southeastern Alaska, which has been closed since 1928. The company determined that the use of "froth-flotation mills" could make it profitable to reopen the site. The froth-flotation technique involves placing crushed rock from the mine into tanks of churning water. Chemicals are added to cause the gold-bearing minerals to rise in the soup, which are then scooped off the surface. The problem is what to do with the slurry—the poisonous waste containing aluminum, copper, lead, and mercury that remains. The slurry's volume consists of 30 percent crushed rock (tailings) and 70 percent water.

Ordinarily, the slurry is disposed of in a tailings pond, where gravity takes over. The tailings settle to the bottom of the pond, and the surface water (which is still poisonous) is reused in the froth-flotation mills. The process can be repeated and repeated, so the wastewater continues to be used in the mining process and is not discharged into the environment. In 1982, the EPA issued a performance standard (through a notice-and-comment rulemaking that resulted in a regulation) for mining operations: new facilities would be held to a "zero discharge" limit. In other words, the EPA prohibited direct discharges of wastewater into waters of the United States from new froth-flotation operations; the EPA would not issue a section 402 permit for such an activity.

Coeur Alaska, however, wanted to discharge the slurry from the Kensington mine directly into Lower Slate Lake, located three miles from Tongass National Forest. In fact, the company wanted to discharge 4.5 million tons of tailings (plus the water) into the lake. If the company had wanted to discharge a small amount—say one ton—then the slurry would have been considered a pollutant subject to EPA jurisdiction and prohibited under the "zero discharge" standard. Ironically, because of the magnitude of the discharge, the company (and the Corps and EPA) considered it to be fill material and thus subject to Corps jurisdiction under section 404. Why was the slurry fill material? Because the effect of the discharge would raise the bottom elevation of the lakebed by 50 feet, and thus under the effects-based test adopted by the Corps in 2002, the slurry was fill material. The Corps decided to grant the section 404 permit.

In addition to raising the bottom elevation of the Lower Slate Lake, the slurry would render it lifeless, which understandably concerned environmental groups. Southeast Alaska Conservation Council (SEACC) and others brought a lawsuit, claiming that the EPA, not the Corps, was the proper permitting agency. SEACC contended that the EPA zero-discharge performance standard controlled. The Ninth Circuit Court of Appeals agreed with SEACC, but Coeur Alaska and the U.S. government sought further review by the Supreme Court.

In a 2009 decision, the Supreme Court held 6-3 that the slurry was fill material, that the Corps was the proper permitting agency, and that the EPA performance standard was not applicable. In its opinion (written by Justice Kennedy), the Court conducted a *Chevron* analysis of the agencies' positions. After determining that the slurry fell under the regulatory definition of fill material, the Court asked whether the EPA performance standard nevertheless applied to fill material. The Court found the Clean Water Act ambiguous on this point. While section 306 made it unlawful for a discharge to violate an EPA performance standard, section 404 seemed to give the Corps full authority to grant permits for fill material without any reference to EPA performance standards. Thus, Congress had not spoken directly to this question, and the Court moved on to the second step of *Chevron*: was the agencies' interpretation of the Clean Water Act reasonable?

But, like the Clean Water Act itself, the Corps' and the EPA's regulations did not expressly address the matter. The Court then turned to how the agencies had interpreted their ambiguous regulations, which of course led the Court to a guidance document. In particular, it examined a 2004 EPA memorandum written by Diane Regas, then director of the EPA's Office of Wetlands, Oceans and Watersheds in Washington, D.C. The Regas memorandum explained that because the discharge of fill material such as slurry does not require a section 402 permit, EPA performance standards would not be applicable. The Court noted that although the Regas memorandum did not warrant *Chevron* deference (it was not subjected to a notice-and-comment process), it still was "entitled to a measure of deference because it interprets the agencies' own regulatory scheme." The Court concluded that the Regas memorandum was "a reasonable interpretation of the regulatory regime," and the Court deferred "to the interpretation because it is not 'plainly erroneous or inconsistent with the regulation[s].'" Even if *Chevron* deference is not applicable, deference is still deference.

Such deference carried over to the Court's review of whether the proposed operation met the section 404 permitting standards. The Court referred to the Corps' statement that even though the slurry will kill all the life in the lake, the reclamation efforts post-mining will result in a lake that "will be at least as environmentally hospitable, if not more so, than now." The Court appeared to accept this statement at face value, which recalls Samuel Johnson's observation regarding a second marriage as "the triumph of hope over experience."

Although in this case the Corps and the EPA ultimately agreed on a consistent definition of fill material, such harmony is not always the norm in the administration of the section 404 program. We now turn to the experience of the Corps in the permit decision process and its odd marriage and complicated relationship with the EPA.

Chapter 5

Strange Bedfellows: The U.S. Environmental Protection Agency and the U.S. Army Corps of Engineers

[S]ection 404 lies like an open wound across the body of environmental law. . . . [It] is constructed on the backs of two beasts moving in different directions.

—*Professor Oliver Houck (1989)*

Ensconced within the Pentagon is the Office of the Assistant Secretary of the Army (Civil Works). This assistant secretary, who is appointed by the president and confirmed by the Senate, has responsibility for overseeing the U.S. Army Corps of Engineers, including its regulatory program, which annually issues permits for tens of thousands of activities affecting wetlands, the vast majority of which are on private property. Which begs the question: why does a military organization have the authority to dictate to civilians—a housing developer, Walmart, even your grandmother—what they can do on their own property?

In 1972, when drafting the Clean Water Act, Congress was divided. The House of Representatives voted to give the section 404 permit program to the Corps, while the Senate preferred the EPA. The House was persuaded by the Corps' long experience with its regulatory program under the Rivers and Harbors Act. The Corps was amenable to accepting this new responsibility in part because it could then maintain permitting control over its own civil works projects, such as flood control and navigation work (Hough and Robertson, 2009). It was precisely this history, however, that

THE FAR SIDE® By GARY LARSON

"Oh, there goes Lenny again—draining off the goldfish bowl. ... He wants to one day work for the Army Corps of Engineers, you know."

FIGURE 5-1.

concerned the Senate. The Corps' reputation was that it never saw a river it did not want to dam; its reputation was not one of environmental sensitivity (figure 5-1). How could such an agency with its construction mission be trusted as the guardians of environmental resources?

In the end, the legislative compromise enshrined in section 404 of the Clean Water Act created one of the most unusual agency relationships in U.S. history. The Corps was given the authority to issue permits for the discharge of dredged or fill material, but in making such decisions, it was required to follow the EPA's regulations. In developing these permit standards, the EPA was required to consult with the Corps. Both agencies were given enforcement powers to take action against people who violated the Clean Water Act. Most remarkable, however, was the ultimate trump card bestowed upon the EPA. If it disagreed with the issuance of a Corps section 404 permit, the EPA could veto the decision. The agencies' different missions and philosophies (protecting the environment versus building dams) led, as Professor Houck vividly suggested, to tension and conflict over the years.

The misnamed 404(b)(1) Guidelines:
More than mere guidance

In Clean Water Act section 404(b)(1), Congress directed the EPA to develop "guidelines" for the Corps to use in making section 404 permit decisions. The use of the term "guidelines" was unfortunate, as it created some initial uncertainty about the binding nature of these standards. In December 1980, in the waning days of the Carter administration, the EPA resolved any ambiguity by completing a public notice-and-comment rulemaking process. The 404(b)(1) Guidelines (named after the subsection calling for their creation) are regulations, codified in the Code of Federal Regulations at 40 C.F.R. part 230.

The 404(b)(1) Guidelines must not be confused with the *Corps'* own permit regulations, which the Corps had issued for its regulatory program under the Rivers and Harbors Act. In 1968, the Corps revised its regulations to include a "public interest review" of potential projects requiring a Rivers and Harbors Act permit. Previously, the Corps' primary consideration was impacts to navigation. Now, as part of the public interest review, the Corps would reject a permit based on its wetland impacts (a new interpretation that was upheld in cases such as *Zabel v. Tabb*[1]). But the public interest review, codified at 33 C.F.R. § 320.4, encompassed a panoply of other factors:

> conservation, economics, aesthetics, general environmental concerns, wetlands, historic properties, fish and wildlife values, flood hazards, floodplain values, land use, navigation, shore erosion and accretion, recreation, water supply and conservation, water quality, energy needs, safety, food and fiber production, mineral needs, considerations of property ownership and, in general, the needs and welfare of the people.

This public interest review has been much maligned, derided as "virtually standardless" (Houck, 1989) and a "parody of standardless administrative choice" (Rodgers, 1994). Although originally developed for its Rivers and Harbors Act program, the Corps also uses it in its Clean Water Act section 404 permit decision-making process. Indeed, in *Rapanos*, Justice Scalia invoked the public interest review when describing the Corps as "an enlightened despot."

As a practical matter, however, the Corps does not deny section 404 permits based solely on the public interest review. A project must pass both the public interest review and the 404(b)(1) Guidelines, and it is the 404(b)(1) Guidelines that are dispositive. If a project fails to meet the 404(b)(1) Guidelines, it will fail the public interest review too, but the latter is not the reason for the permit denial. In a previous publication, I offered to buy lunch to the first person who provided me a copy of a section 404 permit denial in the last ten years that was based solely on the public interest review (Gardner, 2007). I have not had to pay, which suggests that the assumption is valid or the piece was not widely read (probably both). In any event, it is the 404(b)(1) Guidelines that control the section 404 permit decision-making process.

The heart of the Guidelines: The alternatives analysis

The 404(b)(1) Guidelines take up fifty-four pages in the Code of Federal Regulations, but at its heart is the alternatives analysis. The 404(b)(1) Guidelines require permit applicants to demonstrate that there are no "less environmentally damaging practicable alternatives." That awkward phrase can be reduced to a single word: avoidance. Wetlands impacts should be avoided to the extent practicable.

The 404(b)(1) Guidelines establish two rebuttable presumptions to enforce the requirement to avoid wetland impacts. But first, the project purpose must be examined. Is the project water-dependent; that is, must the project be built in or near water? A marina is a water-dependent activity; a housing development is not. For non-water-dependent activities, the 404(b)(1) Guidelines presume that upland sites are available. The 404(b)(1) Guidelines also presume that building the project in these upland sites would be less environmentally damaging than doing so in a wetland site. The permit applicant can rebut these presumptions. Perhaps the project is so large, and it is in an area with so many wetlands, that wetland impacts cannot be avoided entirely. Or perhaps the uplands where the project could be built are more environmentally valuable than a wetland site. Whatever the situation, the permit applicant must explain how it is trying to avoid wetland impacts to the extent practicable. The permit applicant must demonstrate that it has conducted a proper alternatives analysis, considering a range of other options.

These presumptions and the avoidance requirement would seem to preclude many projects from being built in wetlands, but what do they mean in practice? Once again, we must look beyond the statute and regulations, and examine agency guidance. The Corps and the EPA issued a "Memorandum to the Field" in 1993 and then again in 1995 to explain the scope of the alternatives analysis required by the 404(b)(1) Guidelines. The two memoranda did not go through a notice-and-comment process, and thus they are not regulations having the force of law. Rather, they are interpretive rules that clarify how the agencies are reading the regulations (the 404(b)(1) Guidelines), and it is these documents that provide critical insight into how the Clean Water Act section 404 program is actually applied on the ground (figure 5-2).

The 1993 memorandum emphasizes the flexibility inherent in the 404(b)(1) Guidelines. Quoting pertinent sections of the regulation, the memorandum advises field personnel that the alternatives analysis depends on both the quality of the wetland and the type of project: "The level of

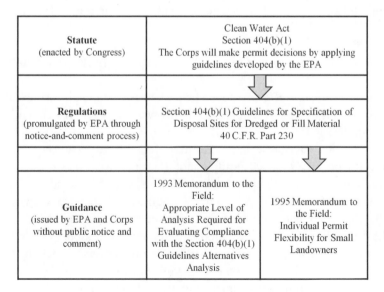

Statute (enacted by Congress)	Clean Water Act Section 404(b)(1) The Corps will make permit decisions by applying guidelines developed by the EPA	
Regulations (promulgated by EPA through notice-and-comment process)	Section 404(b)(1) Guidelines for Specification of Disposal Sites for Dredged or Fill Material 40 C.F.R. Part 230	
Guidance (issued by EPA and Corps without public notice and comment)	1993 Memorandum to the Field: Appropriate Level of Analysis Required for Evaluating Compliance with the Section 404(b)(1) Guidelines Alternatives Analysis	1995 Memorandum to the Field: Individual Permit Flexibility for Small Landowners

FIGURE 5-2. The legal and policy framework for the alternatives analysis.

scrutiny required by the Guidelines is commensurate with the severity of the environmental impact (as determined by the functions of the aquatic resource and the nature of the proposed activity) and the scope/cost of the project." Accordingly, the agencies interpret and apply the requirement to avoid wetland impacts on a sliding scale. If high-value wetlands will be affected, more alternatives must be considered; if it is a degraded wetland that will be filled, fewer alternatives need to be considered. Moreover, as a project's scope and cost increase, so should the range of alternatives that must be analyzed.

The 1995 Memorandum to the Field provides much more specificity about the level of flexibility that may be afforded to small landowners. It states that for projects to construct or expand homes, farm buildings, or small businesses that affect less than 2 acres of nontidal wetlands, the agencies will presume "that alternatives located on property not currently owned by the applicant are not practicable under the Section 404(b)(1) Guidelines." Thus, for these permit applicants the requirement to avoid has been relaxed, if not jettisoned entirely. In effect, the guidance document has announced a presumption (small landowners cannot avoid wetland impacts) to "clarify" the 404(b)(1) Guidelines' presumption (that nonaquatic sites are available). This not only illustrates the ability of agencies to interpret regulations, but it underscores the need to seek out relevant agency guidance documents. One will not find this level of detail in regulations, but it is likely how the Corps regulator will proceed in the field.

Fund for Animals v. Rice: The alternatives analysis in practice

Fund for Animals v. Rice provides a helpful, relatively straightforward alternatives analysis. It also reminds us that courts are generally deferential to agency decisions.

In 1989, Sarasota County, Florida, applied for a section 404 permit to construct an 895-acre landfill. Landfills do not need to be located in or near water; indeed, the potential leaching of waste would advise against such a placement. Because a landfill is a non-water-dependent activity, the 404(b)(1) Guidelines presume that there are upland sites available. Given the size of the project and the amount of wetlands in Sarasota County, however, it was not possible to find a location without adversely affecting some wetlands. Nevertheless, the 404(b)(1) Guidelines still required a search for the "least environmentally damaging practicable alternative."

In its alternatives analysis, Sarasota County identified four potential sites, each of which had a significant amount of wetlands. Site D was 2,130 acres with 19 percent wetlands (404 acres), Site E was 3,360 acres with 22 percent wetlands (739 acres), Site F was 6,150 acres with 22 percent wetlands (1,353 acres), and Site G consisted of 2,100 acres with 13 percent wetlands (273 acres). Because the wetlands on each site were scattered, the wetland impacts in terms of acreage would vary. Placing the landfill in Site D would have the greatest wetland impact (in terms of area) at 92 acres. Site E would affect 61 acres of wetlands, Site F 83 acres, and Site G 78 acres.

At first glance, one might choose Site E as the least environmentally damaging practicable alternative. This site, however, was adjacent to Myakka State Park and drained into the Myakka River. Furthermore, Site E provided habitat for the Florida sandhill crane, a protected species under state law. Endangered species considerations similarly rendered Sites D and G unattractive. Site D had a nesting bald eagle, while Site G likely hosted the eastern indigo snake and was designated as Florida panther habitat. Thus, in the end, the Corps chose Site F (known as the Walton Tract) as the preferred alternative. The Corps also noted that the size of the site (6,150 acres) would result in a large buffer area between the landfill and adjoining properties.

After almost five years of study, in 1994 the Corps issued a section 404 permit to Sarasota County to proceed with a landfill on the Walton Tract. Environmental groups sued on a number of grounds, including a challenge to the alternative analysis. To prevail on that point, the environmental groups would need to show that the Corps had acted "arbitrarily and capriciously" in granting the permit, a very high hurdle. While the environmental groups could not identify an upland alternative, they suggested that the existing Okeechobee landfill could be used by Sarasota County. The court noted that the Corps had considered and rejected that option as impracticable. In addition to questions about the capacity of the Okeechobee landfill, it was located outside of Sarasota County and was not currently accepting intercounty waste. The Corps' permit decision, therefore, was not arbitrary and capricious.

There are several lessons one can draw from this case. First, the permit process for a controversial project such as a landfill (which is always controversial to prospective neighbors) can take years. Second, an advocate would do best to prevail at the agency level. The Corps has a great deal of discretion in whether and under what conditions to issue a permit, and courts will generally be deferential to its judgment. Third, how one defines project purpose will drive the scope of the alternatives analysis.

Defining the project purpose of a golf course:
Jack Nicklaus takes a mulligan.

Although a golf course has water hazards, it is not a water-dependent activity. (Whether those water hazards should count toward achieving the goal of no net loss is a different question that will be taken up in the next chapter.) Golf courses do not need to be constructed in or near water to satisfy the overall project purpose, which is to allow garishly dressed people to spend several hours a day whacking a small ball with a stick. Yet even the design and construction of a golf course can raise questions about the 404(b)(1) Guidelines alternatives analysis.

For example, what is the appropriate geographic range to search for less environmentally damaging sites? In *Stewart v. Potts*, the city of Lake Jackson wanted to build a municipal golf course and restricted its search for sites to within its boundaries. Because a project purpose was to provide a recreational amenity for the residents of the city, the Corps agreed that it was appropriate to consider alternatives only within that particular jurisdiction. Other golf courses may be subjected to a more rigorous alternatives analysis, as Jack Nicklaus discovered.

After winning a record eighteen Grand Slam tournaments, the "Golden Bear" focused his attention on business ventures such as designing golf courses. His "Jack Nicklaus signature" golf courses are challenging—and expensive—to play. They often are accompanied by an upscale housing development, and homeowners pay a premium to be adjacent to these prestigious links. Given the size of the projects (hundreds of acres) and their location (such as in Florida), sometimes the projects required Clean Water Act section 404 permits to fill wetlands. His project in Old Cutler Bay in Florida was particularly controversial.

The project purpose for Old Cutler Bay was defined in a manner to limit the search for practicable alternatives (and thus the need to avoid wetland impacts). Specifically, the project for Old Cutler Bay was described as "an upscale residential/(Jack Nicklaus designed) championship golf course community in South Dade County." Golf courses come in various sizes, from the smaller executive courses (primarily par threes) to the longer championship courses. A Jack Nicklaus signature championship golf course is longer still. The developers stated that the golf course needed to be a particular size or Nicklaus would not approve, and without his imprimatur the adjoining 428 homes would fetch lower prices, thereby rendering the project unprofitable. The Jacksonville District of the Corps accepted this formulation of the project purpose and announced its intention to permit the destruction of approximately 59 acres of wetlands, 47 acres of which were low-value, degraded by the presence of the invasive Brazilian pepper, and 12 acres of which were higher-value white mangroves. The EPA was concerned about the Corps' approach and requested an "elevation," or higher-level review by the Assistant Secretary of the Army (Civil Works) pursuant to Clean Water Act section 404(q).

A 404(q) elevation is essentially an internal governmental appeal by the EPA (or other federal agency), and it provides yet another example of how agencies interpret statutory language. Review the plain language of the Clean Water Act section 404(q): it does not use the term "elevation" or set forth a process for internal reviews. Instead, the focus of the section is "to minimize, to the maximum extent practicable, duplication, needless paperwork, and delays in the issuance of [section 404] permits" and the Corps is called upon to enter into agreements with other federal agencies to ensure that permit applications are speedily processed. The resulting interagency memoranda of agreements (the section 404(q) MOAs) establish a process to resolve interagency disputes in the permit process. If the EPA (or another agency such as the Fish and Wildlife Service) has an objection to the issuance of a section 404 permit, it can invoke the elevation procedures in the section 404(q) MOA and request that the matter be reviewed by Corps headquarters or the Pentagon. While this elevation process may be a reasonable way to resolve disputes, it does not necessarily speed up the permit process as contemplated by the language of section 404(q), which highlights again the importance for one to go beyond the statute and regulations to gain a full understanding of agency practices.[2]

In Old Cutler Bay, the EPA elevated the permit decision based on the Corps' deference to the applicant's narrow project purpose, which included a reference to Jack Nicklaus himself and 428 residential units. The EPA considered this approach an incorrect application of the section 404(b)(1) Guidelines, and if followed elsewhere would result in diminishing the utility of the alternatives analysis in avoiding wetland impacts. Once elevated for higher-level review, Corps headquarters eventually agreed with the EPA, finding that the Corps district had accepted an overly restrictive project purpose (U.S. Army Corps of Engineers, 1990). Instead, the project purpose was characterized as "a viable upscale residential community with an associated regulation golf course in the south Dade County area." The references to Nicklaus and the specific number of units were eliminated, and the Corps district was ordered to revaluate the project.

So the elevation process spared 12 acres of white mangroves, at least temporarily. Critics of the elevation process argue that saving this small area was not worth the time and money spent by the developer and federal government. But the elevation was about more than mere acreage: it was about the larger principle of the proper use of the alternatives analysis to avoid wetland impacts.

Mississippi casinos: Is gambling a water-dependent activity?

Just as the law can call a cornfield a navigable water, it can cause casinos to be considered a water-dependent activity. Sometimes a project's purpose and the resulting alternatives analysis are affected by factors that might seem extraneous. The case of Mississippi casinos offers a prime example.

In the early 1990s, Mississippi state legislators faced a political dilemma. Faced with a budget deficit of hundreds of millions of dollars that ran afoul of a balanced-budget amendment to the state constitution, they wanted to attract businesses to the state to create new jobs and raise revenue. One readily available solution was casinos, and the gaming industry (as it likes to be called) wanted to expand into Mississippi. But many voters in Mississippi are religiously conservative and view gambling as sinful. (While the Bible does not appear to be explicit on whether blackjack and craps are offenses against God, it is susceptible to a number of interpretations. Once again, albeit in a different context, we see the importance of understanding how an implementing organization interprets and applies its rules and regulations [Jacobs, 2007].) Eventually, the lure of raising revenue without raising taxes proved more powerful than scriptural proscriptions or the wrath of conservative voters. As a compromise, however, the state legislators limited the areas where gambling could occur. Roulette and other such vices were prohibited on Mississippi soil, but would be permitted on the state's navigable waters (navigable in the classical sense).

A construction project in a navigable water, including tidal wetlands, requires a Clean Water Act section 404 permit. When the casino developers sought these permits, the first question the Corps needed to answer was whether casinos and related structures, such as hotels and parking lots, were water-dependent activities. Ordinarily, one would conclude that shooting dice, doubling down, and letting it ride do not require the presence of water. Certainly hotels and parking lots generally do not need to be located in or near water to accomplish their purposes of housing guests and their vehicles. Thus, the presumptions of the section 404(b)(1) Guidelines would kick in—that upland sites were available, and these sites were less environmentally damaging practicable alternatives.

Presumptions, however, are just a starting point, and they can be overridden or rebutted. Here the casino developers pointed to the state law requiring that casinos be located only in navigable waters. Thus, a gambling operation in Mississippi is a water-dependent activity *by virtue of state law*. As such, there is no need to avoid wetland impacts entirely. A casino developer still had to demonstrate that its site was the least damaging practicable alternative among the options available, which could be accomplished by avoiding high-quality wetland areas. Moreover, the 404(b)(1) Guidelines require permittees to minimize any unavoidable impacts (perhaps by timing construction to avoid sensitive nesting or spawning seasons) and to compensate for any remaining environmental impacts through restoration, enhancement, creation, and/or preservation of other wetlands. Nevertheless, casinos and their related structures sprouted along the coastline.

The Corps' Vicksburg district was responsible for deciding whether to issue section 404 permits for the casinos. When the Corps issued permit after permit, the EPA raised concerns that the Corps was not properly applying the 404(b)(1) Guidelines. The Office of the Assistant Secretary of the Army (Civil Works) agreed and issued a moratorium on further casino permits until the cumulative impacts of the industry could be studied. Casino backers, exercising their First Amendment right to petition the government for a redress of grievances, complained to U.S. Senator Trent Lott. He responded by holding up the confirmation of unrelated EPA political appointees and raising conflict-of-interest claims against the government employees who were delaying casino development (Weissman et al., 1999). After an internal review, the Pentagon inspector general concluded that the conflict of interest charges were unfounded. Nevertheless, Senator Lott's pressure was effective. The Corps and the EPA backed off, and casino projects continued along the coast relatively unabated. By 2005, Mississippi had thirteen casinos along the Gulf and more upstream on the Mississippi River. Annual gross revenues were averaging more than $2.5 billion, third highest in the country (behind Nevada and New Jersey), and casinos provided at least 17,000 local jobs.

Until Hurricane Katrina. While much of the focus in Katrina's aftermath was the inundation of New Orleans, a large swath of the Gulf Coast including Mississippi and Alabama was also devastated. All of Mississippi's coastal casinos, along with Senator Lott's home in Pascagoula, were destroyed (figure 5-3).

In an effort to assist casinos with rebuilding efforts (to bring back jobs,

FIGURE 5-3. A casino on top of a hotel post-Katrina. (Source: U.S. Geological Survey.)

visitors, and revenues), the Mississippi state legislature eased the restrictions on the location of casinos. They may now be built up to 800 feet inland from a navigable water. Accordingly, casinos in Mississippi should probably no longer be considered a water-dependent activity. In theory, if a proposed casino project would require the filling of wetlands, the presumptions of the 404(b)(1) Guidelines would now apply, making it more challenging to receive a permit. In reality, however, it is not likely that the Corps will deny permits as Mississippi communities begin to rebuild. It is also unlikely that the EPA will veto these permit decisions.

Sweedens Swamp and the market-entry theory: "It depends on what the meaning of the word 'is' is."

Long before President Clinton demonstrated his ability to parse the English language,[3] the 1986 veto to protect Sweedens Swamp illustrated that the word "is" can be interpreted in several ways. Under Section 404(c), the EPA may veto any Corps permit that results in unacceptable adverse environmental effects. The Corps issues thousands of individual permits annually, and the veto, although a potential hammer, has been employed sparingly. On occasion, however, the EPA has used the veto to vindicate its interpretation of the 404(b)(1) Guidelines. It did so in the case of Sweedens Swamp to uphold the "market-entry" approach in the alternatives analysis.

The alternatives analysis requires that "practicable" alternatives be examined—but what does "practicable" mean? The 404(b)(1) Guidelines specifies that an alternative is practicable when "it *is available* and capable of being done, after taking into consideration cost, existing technology, and logistics in light of overall project purposes" (emphasis added). While this regulatory definition clarifies the term a bit, it creates additional questions. When, for example, is a site available?

A dispute between the Corps and the EPA over the issue of availability came to a head over a proposed shopping mall in Attleboro, Massachusetts. In April 1982, Pyramid, a developer, purchased an 80-acre site in Attleboro with the intent to construct a shopping mall. While the Attleboro site was well situated for a mall, unfortunately for Pyramid it was also the location of Sweedens Swamp, which consisted of almost 50 acres of red maple swamp. Three miles away in North Attleboro, an alternative upland site was available; however, a competing mall developer later acquired a purchase option on the North Attleboro site in July 1983. The following summer Pyramid applied for a section 404 permit to fill in Sweedens Swamp, and one of the key questions was at what point does the alternatives analysis begin—at the time that Pyramid entered the market to construct a mall or at the time of the permit application? Under the market-entry test, the North Attleboro site (although now in the control of a competitor) could be considered a practicable alternative. Under the time-of-application test, the North Attleboro site would not be evaluated.

The Corps adopted the time-of-application test, which eliminated the need to consider the North Attleboro (upland) site. The EPA disagreed and endorsed the market-entry approach. In vetoing the project, the EPA concluded that the shopping mall would result in unacceptable impacts to wildlife, in part because of the availability (in the past) of an upland site— the North Attleboro property. Of course, that site was no longer available, and Pyramid sued, claiming that the EPA's interpretation of its regulations was arbitrary and capricious.

Pyramid observed that the 404(b)(1) Guidelines use the present tense—"is available"—and contended that the EPA's market-entry test essentially changed "is" to "was." Pyramid also relied on the U.S. Supreme Court's decision in *Gwaltney* (discussed in chapter 2) as precedent for the importance of paying attention to verb tenses. The agency responded that it was interpreting "availability" in light of the overall purpose of the Clean Water Act and the 404(b)(1) Guidelines, which is to encourage developers to avoid wetland impacts. In the EPA's view, the market-entry test was necessary to ensure that developers did not game the system, waiting to apply for a permit after upland alternatives were no longer available. The case eventually reached the Second Circuit Court of Appeals, which upheld the EPA's interpretation as reasonable in a 2-1 decision. The dissenting judge pointed to the present tense ("is available") and noted that the market-entry test was inherently vague:

> When does a developer enter the market? When he first contemplates a development in the area? If so, in what area—the neighborhood, the village, the town, the state or the region? Does he enter the market when he first takes some affirmative action? If so, is that when he instructs his staff to research possible sites, when he commits money for more intensive study of those sites, when he contacts a real estate broker, when he first visits a site, or when he makes his first offer to purchase?

Nevertheless, the EPA emerged with a victory and a mandate to require that the Corps use the market-entry test when conducting an alternatives analysis.

This case, known as *Bersani v. Environmental Protection Agency*, is included in many environmental law casebooks. But here's the funny thing: generally the Corps does not use a market-entry test in the field. It may have the authority to do so, but as a practical matter it *is* difficult to apply. So *Bersani* may be interesting from an academic perspective, but does not provide realistic insights into the day-to-day application of the Clean Water Act section 404 program. Court decisions might tell us what the outer limits of an agency's authority are, but they do not necessarily tell us what is happening in the field.

The mitigation MOA: Resolving the buy-down and sequencing dispute

Another central issue in *Bersani* (and one that tends to be ignored in the environmental law course books) was the role of compensatory mitigation in the permit process. Pyramid proposed to fill 32 acres of Sweedens Swamp, but promised to create 36 acres of wetlands elsewhere. In the Corps' view, the compensatory mitigation could be considered in the alternatives analysis. If a permit applicant offered an attractive compensatory mitigation package, thus reducing overall impacts to a negligible level, there was no need to conduct a rigorous alternatives analysis. The permit applicant could in effect "buy down" the impacts through compensatory mitigation and thereby avoid the avoidance requirement (Gardner, 1990).

Even though the Corps of Engineers is nominally a military organization, its regulatory program operates in a decentralized fashion. There are thirty-eight Corps districts in the United States, and there are variations between (and even within) each district with respect to interpreting regulations and guidance. In the 1980s, some Corps districts used the buy-down approach. Other districts, however, advocated using a sequential approach: avoid impacts, then minimize unavoidable impacts, and finally compensate for any remaining impacts through restoration, enhancement, creation, and/or preservation of other wetlands. Under the mitigation "sequence," compensatory mitigation could only be considered at the end of the process. An attractive mitigation package could not be employed to reduce or eliminate the alternatives analysis and the initial objective of avoiding wetland impacts. The EPA interpreted the 404(b)(1) Guidelines to require the "avoid-minimize-compensate" sequence.

The inconsistent application of the 404(b)(1) Guidelines did not go unnoticed. In 1986, the Senate Subcommittee on Environmental Pollution directed the two agencies to form a work group to respond to concerns from the environmental community, private landowners, and industry groups. After years of discussions, the Corps and the EPA settled their differences in a memorandum of agreement (MOA), which was published in the *Federal Register* in March 1990. The MOA contained interpretive rules and statements of policy, and the agencies did not seek public notice and comment. The MOA was not a regulation and did not have the force of law. Nevertheless, it contained very important information and guidance about how the 404(b)(1) Guidelines (which are regulations with the force of law) would be applied by regulators in the field.

The MOA committed the Corps for the first time to an overall goal of "no net loss" of wetlands within the section 404 regulatory program. This statement of policy was clearly aspirational, and the MOA recognized that it may not be feasible to achieve no net loss through compensatory mitigation in each and every permit action. But the MOA did note that compensatory mitigation would play an important role in striving for no net loss. Even more important, the MOA clarified the appropriate procedures for considering compensatory mitigation; it endorsed the EPA's approach.

Statute Clean Water Act (enacted by Congress)	33 U.S.C. § 1344	Section 404(b)(1): The Corps will make permit decisions by applying guidelines developed by the EPA
Regulations (promulgated by EPA through notice-and-comment process)	40 C.F.R. Part 230	Section 230.10 (a): No permit shall be granted "if there is a practicable alternative to the proposed discharge which would have less adverse impact on the aquatic ecosystem." Section 230.10(d): No permit shall be granted "unless appropriate and practicable steps have been taken which will minimize potential adverse impacts on the aquatic ecosystem." Section 230.75(d): "Habitat development and restoration techniques can be used to minimize adverse impacts and to compensate for destroyed habitat."
Guidance (issued by EPA and Corps without public notice and comment)	1990 Mitigation MOA	The Corps should evaluate applications for individual permits' through the following sequence: • Avoid impacts to aquatic resources if a less environmentally damaging practicable alternative exists • Minimize unavoidable impacts by taking appropriate and practicable steps • Compensate for any remaining impacts through restoration, enhancement, creation and/or preservation of other aquatic resources

FIGURE 5-4. The legal and policy framework for the "avoid-minimize-compensate" sequence.

Henceforth, all Corps districts would apply the sequence of avoid, minimize, and compensate (figure 5-4). Compensatory mitigation could not be used to buy down a project's impacts to bypass the avoidance step (i.e., the alternatives analysis). Although the MOA was not a legally binding document, it nevertheless had a significant impact on the way that the Corps reviewed permit applications.

When the 1990 MOA was published, environmental groups and the regulated community both were critical (Gardner, 1990). Environmental groups were outraged at the process. They were not concerned that the public played no role; rather, they were concerned that after the Corps and the EPA had negotiated the MOA, it was reviewed and revised by John Sununu, then–White House Chief of Staff, and others in the Bush administration. The environmental groups claimed that Sununu "pulled rank" and had watered down the MOA, making certain provisions ambiguous. Putting aside the issue of whether the slight changes eviscerated the environmental benefits of the document,[4] it was entirely appropriate for the White House to review the MOA. The Corps and the EPA are executive branch agencies, and their ultimate boss is the president.

Furthermore, the final MOA could not have been entirely toothless, because the regulated community sued to invalidate it. ARCO Alaska, Inc. and others claimed that the MOA should have gone through a notice-and-comment process. Although these groups could establish standing to sue (the alleged procedural error led to an MOA that would injure them by increasing the costs of projects), the Ninth Circuit Court of Appeals ruled that their lawsuit was brought too soon (*Municipality of Anchorage v. United States*, 1992). Their claims were not yet *ripe*, in part because of the ambiguities contained in the MOA. Accordingly, any such challenge to the MOA would need to be done in the context of its application to a specific project. Ironically, the changes that the environmental groups criticized actually served to insulate the MOA from legal attacks.

The old Corps returns: The EPA vetoes the Yazoo River Project

While the 1990 MOA illustrates that the Corps and the EPA can work together (at least at the Washington level), the cooperative nature of the relationship has varied over the years. Sometimes interagency tensions resulted from personality conflicts, and sometimes they arose from genuine policy differences (or a combination of the two). The Corps' Yazoo Backwater Area Project, and the EPA's initiation of its section 404(c) veto process in 2008, demonstrates that despite progress, the historic missions of the two agencies can still work at cross-purposes.

Located in the Yazoo River Basin in west-central Mississippi, the Yazoo Backwater Area is prone to frequent flooding (figure 5-5). As water levels reach a certain point in the Mississippi River, its waters back up into its

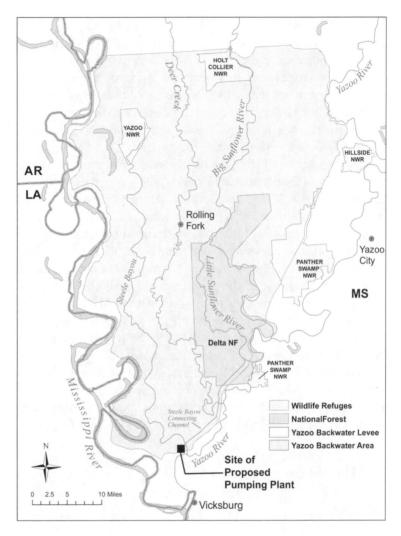

FIGURE 5-5. Location of the Yazoo Backwater Area Project. (Source: U.S. Fish and Wildlife Service.)

tributaries, including the Yazoo River. When the Yazoo River reaches its capacity, the Yazoo River Basin can no longer be drained, and flooding results. To prevent this flooding, the Corps constructed a floodgate to keep a rising Mississippi River's waters from the Yazoo River. Like many government projects, there was an unintended consequence. When the floodgate is closed, the Yazoo River Basin cannot drain, and flooding still results.

Consequently, in the Flood Control Act of 1941, Congress authorized the Yazoo Backwater Area Project. There were many separate parts to the project. Along with levees and drainage structures, the federal government would construct pumping stations; the federal government (i.e., taxpayers) would bear the entire cost of the project.

Although other federal agencies raised concerns about the impact on wildlife (the U.S. Fish and Wildlife Service did so starting in 1956) and wetlands (as the EPA did in 1982), the Corps proceeded with various pieces of the project as Congress provided appropriations. The Water Resources Development Act of 1986 changed the rules of the game, and required that local sponsors cost-share—that is, pay part of the project costs, 25 percent in this case. No entity in Mississippi wished to contribute this amount, and new work ceased for ten years. Then in 1996 Congress reversed itself and committed the federal government to fully funding the project. The Vicksburg District of the Corps of Engineers went back to work.

When Congress first authorized the project, there were no federal environmental laws to speak of, and wetlands were viewed as nuisances to be drained. Beginning in the 1970s, however, Congress began to enact meaningful environmental laws, such as the National Environmental Policy Act (which requires an environmental study before a federal project can proceed) and the Clean Water Act. When the Vicksburg District recommenced its work in 1996, it needed to comply with these environmental statutes, which caused further delays.

One might think it odd that a project such as the Yazoo Backwater Area Project, which Congress had specifically authorized, could be held up by environmental laws. If Congress approved a project (and funded it), shouldn't the Corps proceed? But the National Environmental Policy Act and the Clean Water Act were also passed by Congress. A typical approach to statutory interpretation is to try to reconcile laws that seemingly are in conflict with each other. In this context, the Corps may move forward with the project, but it must do so in accordance with environmental statutes. If Congress desires that a project be exempt from environmental statutes and regulations, it can pass another law declaring the exemption. (Famous, or infamous, examples include the Tellico Dam project on the Little Tennessee River and the border fence in Southern California.) But Congress has not exempted the Yazoo Backwater Area Project from environmental requirements, including the 404(b)(1) Guidelines.

As part of the project, the Corps intended to construct a pumping station that would have the capability of moving water at the rate of 14,000 cubic feet per second, which is the equivalent of more than 9 *billion* gallons per day. While the footprint of the pumping station itself would result in only 52.6 acres of wetland impacts, the environmental impact of the pumping station's operation would be staggering: approximately 67,000 acres of bottomland hardwoods would be adversely affected. More than a third of this area (26,300 acres) would be drained to such an extent that it would no longer be considered wetlands, and the rest (40,700 acres) would be degraded. (Note that 67,000 acres is 104.7 square miles, which is roughly the size of Jackson, Mississippi.) The project was estimated to cost more than $220 million to construct and $15 million annually to operate. The vast scope of the project and its massive environmental impacts harkened back to the Corps of Engineers of yesteryear, in which it did not see a river it did not want to dam or control.

But times and, more important, environmental laws have changed.[5] In March 2008, the EPA announced its intention to veto the Yazoo Backwater Area Project under Clean Water Act 404(c), based on unacceptable environmental impacts, and it indeed vetoed the project in September 2008. This is only the twelfth time that the EPA has exercised its veto authority, and the first time in more than seventeen years. While the 404(c) veto might seem to be a vestige of an earlier era, the Yazoo Backwater Area Project demonstrates its continued relevance. Occasionally, the EPA and the Corps (or at least some Corps districts) are still beasts moving in different directions.

To put the Yazoo Backwater Area Project in perspective, consider that each year from 2000 to 2006 the Corps granted permits to fill between 13,887 and 24,650 acres. The Yazoo project, with its 67,000 acres of impacts, would by itself exceed these annual national wetland losses. The project would seem to be utterly inconsistent with the goal of no net loss, to which we now return.

Chapter 6

No Net Loss: Lies, Damned Lies, and Statistics

The government began formulating agricultural policy in 1794, when the residents of western Pennsylvania started the Whiskey Rebellion in response to an excise tax on corn liquor. The agricultural policy formulated in 1794 was to shoot farmers. In this case, the federal government may have had it right the first time.

—*P. J. O'Rourke*, Parliament of Whores *(1991)*

While P. J. O'Rourke, recalling the federal government's response to the Whiskey Rebellion, directed his invective against farmers in the context of modern-day agricultural subsidies, a similar sentiment could be adopted by wetland activists. Agricultural activities have been the primary cause of historical wetland losses in the United States. But farmers did not act on their own; here the federal and state governments were their partners. Indeed, through a series of Swamp Lands Acts beginning in 1849, the federal government encouraged states to reclaim wetlands (areas "wet and unfit for cultivation") and put them to more productive use to feed a growing nation. It was not until much later that the federal government, public, and even farmers recognized the value of wetlands in their natural state.

The general policy toward wetlands today is captured in the pithy slogan "no net loss." The first Bush administration endorsed a "no net loss" policy, as did the Clinton administration. Like most slogans, it is an oversimplification of a complicated subject. In 2006, outgoing Secretary of the

Interior Gale Norton declared that we had accomplished the objective of no net loss in terms of acreage through various incentive and regulatory programs (Dahl, 2006). Critics ridiculed this assertion of "mission accomplished," noting that golf course water hazards were included in the wetland totals (Barringer, 2006; Colbert Report, 2006). Yet both the (second) Bush administration and its environmental critics were correct.

This chapter takes a brief frolic and detour from our general focus on the Clean Water Act. In examining the history of wetland losses in the United States and the progress, or the lack thereof, we have made toward satisfying the "no net loss" of wetlands policy, it is important to look beyond one regulatory program. We begin by reviewing the past and present threats to wetlands: agricultural activities, general development (such as home-building, road-building, and other construction projects), and exotic species—both plant and animal. We will then consider the government's response to these threats, which includes not only the Clean Water Act section 404 regulatory program, but financial disincentive and incentive programs for farmers and encouraging Cajuns to eat nutria (it's the "other" other white meat). Finally, we will review the effectiveness of these measures. While it is accurate, in one sense, to declare that the United States is meeting the goal of no net loss, it is a hollow victory when one looks beyond the mere acreage totals. And, as we shall see, it is also a bit unfair to blame only the farmers.

The starting point: A nation of farmers

To gauge wetland losses (or gains), you must start with a baseline. The FWS estimates that in the 1780s, the area that makes up the current coterminous United States (i.e., excluding Alaska and Hawaii) had approximately 221 million acres of wetlands of various types (Dahl, 1990). There are several ways to make this estimate, such as reviewing historical agricultural drainage records or using hydric soil data. Soils can maintain hydric characteristics long after they are drained, and thus current records can be relied upon to determine likely past wetland conditions. (And, as we saw in chapter 3, these relic hydric soils should not be the sole criterion to determine whether a site is a wetland today.) While no one knows for certain how many acres of wetlands existed at the time of European settlement, most estimates are close, ranging from 211 million to 221 million acres. What we do know for certain is that we have lost millions of acres since the 1780s.

A frequently cited statistic is that we have lost more than half of our original wetland base. A 1990 FWS report pegged the number at 53 percent, which would be 60 acres an hour, every hour, every day, every year, for 200 years. Note that these statistics exclude consideration of Alaska. Given its size, small population, and relatively undeveloped condition, its inclusion would skew the numbers (and the policy debate). Alaska alone has approximately 170 million acres of wetlands, and less than 1 percent has been developed. If one includes Alaska, then the rate of historic wetland losses for the nation drops to about 30 percent. So one could argue that less than half our wetland base remains (excluding Alaska), or one could contend that more than 70 percent is still with us (including Alaska). The former is the more persuasive case, especially when one begins to look at individual states and regions.

Ten states have lost more than 70 percent of their historic wetland base; California and Ohio have lost more than 90 percent (Dahl and Johnson, 1991). Six states (Illinois, Indiana, Michigan, Minnesota, Ohio, and Wisconsin) account for more than 36 million acres of wetland losses. Florida is the leader in terms of total acreage lost, with more than 9.3 million. What do these states have in common? They are farming states.

It should be no surprise that states with many wetlands would become agricultural hubs. Wetland soils can be extraordinarily fertile. If a wetland can be drained sufficiently or if the water regime can be controlled, a farmer can then reap the benefits of the rich soil (Vileisis, 1997). For example, the Everglades Agricultural Area, south of Lake Okeechobee, has been one of the most productive farming areas in the region.

With government subsidies and price supports, farmers continued to drain and convert wetlands well into the twentieth century. Between 1950 and 1970, annual wetland losses attributable to agricultural activities were about 250,000 acres. By the 1970s and 1980s, the rate of loss was still 290,000 acres per year (Dahl and Johnson, 1991). Only recently, due to agricultural reforms that will be discussed shortly, has farming ceased to be the primary cause of wetland losses. Urban and rural development have taken the lead (Dahl, 2006).

If you build it, they will come.

As the U.S. population increased, more land was brought into agricultural production to satisfy the nation's nutritional needs. But the expanding population also needed somewhere to live. And to shop. Accordingly, wetlands, which often were less expensive than uplands, were ditched and drained for housing developments and shopping centers.

Many people like to live near nature. Unfortunately, this is not a good thing for nature (including wetlands and the species that depend on this habitat). One of the ironies of the home-building industry is that the names of the communities often reflect the habitat destroyed or animals displaced. Thus, the upscale home development called Olde Cypress in Naples, Florida, which wiped out more than 174 acres of wetlands, including cypress sloughs, has inviting neighborhoods such as Egret Cove, Ibis Landing, and Wild Orchid at Woodsedge.

With the advent of large shopping malls and big-box stores (along with their acres of parking lots), wetlands have come under a new threat. An old joke about wetland indicators illustrates the perceived proclivity of developers to choose wetlands as the preferred building site: "How can you tell if an area is a wetland? Look for hydrology, hydrophytic vegetation, hydric soils, and whether there is a Walmart store."

Road construction contributes to wetland losses directly and indirectly. Often these linear projects cannot avoid wetlands. The placement of the fill for the road obviously destroys a particular site, but the impacts are much farther reaching. A road can alter water flows, thereby degrading a larger area of wetlands. For example, I-75, known as Alligator Alley, bisects southern Florida and disrupts the natural sheet flow toward Everglades National Park and Florida Bay. A road can also open up previously inaccessible places to development. Environmental groups opposed the construction of the Suncoast Parkway in west-central Florida not only for its direct impacts on hundreds of acres of wetlands, but for the indirect impacts as well (*Sierra Club v. U.S. Army Corps of Engineers*, 2002). The Suncoast Parkway has contributed to the development of Pasco County (one of the fastest growing counties in the country) and the reduction of its wetland resources.

A growing population also requires a ready energy supply. The search for oil and gas off of Louisiana's coast has contributed to the rapid loss of coastal marshes, and the 2010 Deepwater Horizon oil spill debacle may exacerbate that decline. Mountaintop removal mining in Appalachia, discussed in chapter 4, which involves accessing the coal by removing the tops of mountains and discarding the "overburden" into valleys, has resulted in the destruction of thousands of acres of wetlands and thousands of miles of streams. Once the coal or oil or gas is extracted and used to produce electricity, that energy must be transported to individual homes and businesses. The utility lines that transect our communities also affect wetlands.

Some enterprising attorneys filed a multimillion-dollar lawsuit claiming that the oil companies are responsible for damages related to Hurricane Katrina. Scientists have contended that the oil companies' system of canals degraded Louisiana's coastal marshes (Streever, 2001), which serve as a natural barrier to ameliorate storm surges and flooding. The legal argument is that if the companies' actions had not harmed the coastal wetlands, the wrath of Hurricane Katrina would have been significantly reduced. While there is some causal connection (albeit likely too attenuated to establish tort liability), the oil companies are not solely responsible. Everyone who relies on the energy they produce is implicated.

The other illegal alien problem: Invasive species

It is not just people who are the nonnative species moving into wetlands. For example, Everglades National Park is now home to the Burmese python (Harvey et al., 2008). As its name suggests, the python is not native to Florida, but the Everglades has become a dumping ground for pet owners who tire of their reticulated constrictors. Many people were amused (or grossed out) by the 2005 photograph depicting a fight between a 6-foot Florida alligator and an overly ambitious 13-foot Burmese python (figure 6-1). Neither survived (as the python was split open after attempting to ingest the gator), and clashes such as these graphically illustrate the battle over alien and invasive species (figure 6-2 shows a different encounter).

An alien species is a nonindigenous or nonnative species. Also known as exotic species, these are plants or animals that have been introduced, intentionally or unintentionally, into a new environment. If their new environment does not contain a natural predator or some other form to check the alien species, the newcomer may expand its range, displace native species, and be classified as an invasive or nuisance species. Alien and invasive species can severely degrade wetlands.

Sometimes a species is introduced precisely *because* it can degrade wetlands. For example, John Gifford, the first American to earn a doctorate in forestry, was among those who brought melaleuca, an Australian broad-leaved paperbark tree, to Florida to help drain the Everglades. The Corps of Engineers also used it to prevent flooding and to stabilize shorelines along Lake Okeechobee (Dray et al., 2006). Melaleuca has proven to be too

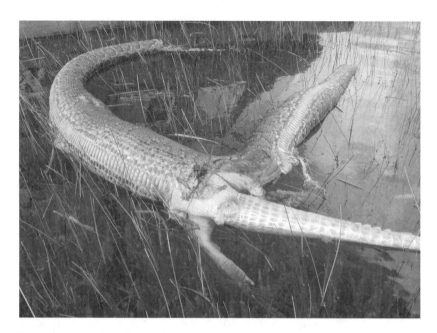

FIGURE 6-1. Duel to the death. (Source: Everglades National Park. Photo credit: Michael Barron.)

FIGURE 6-2. Direct competition. (Source: Everglades National Park. Photo credit: Lori Oberhofer.)

effective. It now occupies more than a half million acres in South Florida, and it is not a good neighbor. Melaleuca takes over cypress swamps, hardwood bottomlands, freshwater marshes, and other habitats. Melaleuca stands can be dense; researchers have reported finding up to 31,000 trees and saplings on a single acre (Mazzotti et al., 1997). The effect is to eliminate native vegetation and animals. Accordingly, melaleuca is now viewed as an invasive species, and the South Florida Water Management District has spent more than $40 million over the past two decades to halt its attack on wetlands. This figure does not include the millions of dollars the federal government has expended for melaleuca control in Everglades National Park and Big Cypress National Preserve.

Sometimes a species is introduced because it is pretty. Consider the case of water hyacinths. A floating aquatic plant native to South America, it was brought to the United States for ornament. It has lovely, bluish lavender flowers. It also grows at an alarming rate and can choke waterways. Its thick mats block sunlight, deplete oxygen levels in waterbodies, and even interfere with navigation. Some states now prohibit its planting, and millions of dollars per year are devoted to battling its spread. Florida alone spends over $3 million per year (Mullin et al., 2000).

Animals can also qualify as invasive species. Perhaps the invasive animal species with the most colorful history is the nutria. A semiaquatic rodent from South America, it was imported to the United States in the 1930s for its fur. E. A. McIlhenny, of Louisiana and Tabasco sauce fame, was among those who owned several dozen nutria. Legend has it that a hurricane destroyed their pens, releasing the nutria into the bayous of Louisiana, where they became fruitful and multiplied. And multiplied, and multiplied again. It seems that nutria's predators, such as alligators, could not keep them in check, and marsh plants offer abundant food. But as they munch away with their orange teeth on the roots of these plants, they convert the marsh into open water. The McIlhenny company historian disputes the details of the story: his research indicates that E. A. McIlhenny did not import nutria from Argentina; rather, he bought them from a preexisting nutria farm in Louisiana (Bernard, 2010). A hurricane did not free the nutria, but McIlhenny did so intentionally (which does not exactly acquit the old sauceman). But what is not disputed is the nutria's range today, which reaches New Jersey, nor its environmental impact. Nutria have been responsible for tens of thousands of acres of wetland losses.

Whether or not the McIlhenny anecdote is true, the McIlhenny family at least had a sense of humor. In 1988 John S. McIlhenny established a charitable organization, the Coypu Foundation, which generously provides grants to higher education. *Coypu* is what some in the Spanish-speaking world call nutria.[1]

With wetlands under assault from farmers, developers, and semi-aquatic rodents with orange teeth, the government (federal and state) has stepped in with a variety of mechanisms to try to stem the losses.

Agricultural sticks and carrots: Swampbuster and the Wetlands Reserve Program

To respond to wetland losses due to agricultural activities, the federal government decided to hit farmers where it would hurt the most: their subsidies. Since the Great Depression, the federal government has offered farmers (and now agribusiness) a suite of economic incentives to ensure the solvency of farming operations and a secure food source. In 1985, however, Congress enacted what is known as the "Swampbuster" program, which represented a dramatic shift in federal policy (Williams, 2005). Instead of encouraging and underwriting farmers' draining of wetlands, the federal government would now penalize such conversion activities. Under Swampbuster, if a farmer drained or altered a wetland to produce an agricultural commodity, the farmer would be ineligible to receive federal benefits, such as loans, subsidized insurance, and price and income supports. Swampbuster is an odd name, and slightly misleading: swamps (and other wetlands) were not being busted; they were being protected. Or at least the federal government was creating disincentives to their draining. Perhaps a more accurate sobriquet would have been "Subsidybuster," but that could have roused greater opposition in the farm lobby.

If the Swampbuster program was a stick, the Wetlands Reserve Program (and other similar programs such as the Conservation Reserve Program and the Environmental Quality Incentives Program) was the carrot. The federal government would pay farmers to undertake environmentally beneficial actions. The farmers would not be raising crops; they would be restoring and protecting wetlands.

Under the Wetlands Reserve Program, the federal government will pay for all or most of the wetland restoration costs. In addition, the government will compensate the farmer in return for the promise that the site will be maintained as a wetland. The level of payments that a farmer receives is contingent on what the government (and public) obtains in return. If the farmer agrees to convey a permanent easement (i.e., promises that the restored area will never be developed or converted back to crops) to the Department of Agriculture, the agency will pay for the easement and 100 percent of restoration costs. For a thirty-year easement (i.e., the restored site cannot be developed or put into agricultural production for thirty years), the agency will provide between 50 and 75 percent of what it would pay for a permanent easement and 75 percent of restoration costs. If the farmer does not wish to convey a property interest such as an easement, the farmer and the agency may agree that the site will be restored and maintained for a minimum of ten years. In such cases, the agency will not make an easement payment, but will pay up to 75 percent of restoration costs.

Established in 1990, the Wetlands Reserve Program has been popular with farmers. From fiscal year 1992 to fiscal year 2007, the Natural Resources Conservation Service (the agency within the Department of Agriculture that administers the Wetlands Reserve Program) entered into more than 10,000 contracts with landowners, enrolling almost 2 million acres at a cost of approximately $2.1 billion (NRCS, 2010). While it is yet another subsidy for the agricultural community, it is a nonregulatory approach to wetland protection and one that provides wide public benefits, such as improved water quality and increased bird habitat. The Wetlands Reserve Program relies on farmers to volunteer, and in light of the burgeoning ethanol and biofuel market, it is an open question whether farmers will continue to enroll land at the current levels.

Offsetting development impacts: Compensatory mitigation

Whereas the Wetlands Reserve Program relies on farmers to volunteer and sign up to restore wetlands, the Clean Water Act takes a more traditional, regulatory approach. A developer (or anyone else) who wishes to fill in a wetland must obtain a section 404 permit from the Corps of Engineers. As explained in chapter 5, the Corps will analyze the proposed project through a sequence: avoid, minimize, and compensate. Wetland impacts must be avoided to the extent practicable (as determined by the alternatives analysis), and any unavoidable impacts must then be minimized. If any wetland impacts remain after minimization, the Corps will require compensatory mitigation. The permittee must promise to restore, enhance, create, and/or preserve other wetlands to offset the environmental impacts of the development project.

The Clean Water Act does not use the term "mitigation." Nor initially did the 404(b)(1) Guidelines. The details of the requirements for compensatory mitigation were found (until very recently) only in guidance documents, such as the 1990 MOA between the EPA and the Corps (National Research Council [hereafter NRC], 2001).

Recall that the 1990 MOA resolved the controversy of buy-downs versus sequencing in favor of the latter: compensatory mitigation could not be used to weaken the alternatives analysis. The 1990 MOA also provided additional guidance on the use of compensatory mitigation. The agencies expressed a preference for on-site mitigation, which is mitigation on or near the development site. The agencies also endorsed in-kind mitigation. In-kind mitigation relates to the type of wetland function that a development project affects. For example, if a housing development destroys wetlands that provided habitat for migratory waterfowl, in-kind mitigation would seek to offset the loss of that particular wetland function. Furthermore, the 1990 MOA favored restoration and enhancement, rather than creation or preservation, as acceptable forms of compensatory mitigation. As justification for this policy, the 1990 MOA noted the uncertainty surrounding the likelihood of success in creating wetlands and noted that preservation of existing wetlands alone does not contribute to no net loss. If 10 acres of filled wetlands are exchanged for 10, 20, or even 30 preserved acres (i.e., legally protected through the use of a conservation easement), there is still a net loss of 10 acres. The agencies therefore counselled that preservation should be used as the sole source of compensatory mitigation only in exceptional cases.

In theory, to achieve the goal of no net loss (and the general goals of the Clean Water Act), the Corps could deny many section 404 permit applications. In practice, it rarely does. One newspaper study (provocatively and accurately titled "They won't say no") found that the Jacksonville District denied just one permit application from 1999 to 2003. During that time period, the district authorized more than 12,000 permits (Pittman and Waite, 2005). As the Corps' own data illustrate, the Jacksonville District's record is not an anomaly. Nationwide, the Corps grants thousands of permits each year and denies only a few hundred (figure 6-3).

Instead of permit denials, the Corps relies heavily on the concept of compensatory mitigation to meet the no net loss goal (Gardner, 2005). We will see shortly that this reliance is unwarranted.

The Cajun solution: Eat a nutria, save a wetland.

Before examining the effectiveness of Swampbuster, the Wetlands Reserve Program, and the Clean Water Act in contributing to the goal of no net

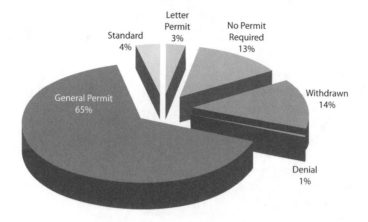

FIGURE 6-3. Corps permit statistics for fiscal years 2007 and 2008. (Source: U.S. Army Corps of Engineers.)

loss, let us consider the government's response to the problem of invasive species. The federal and state governments can and do prohibit the importation or release of invasive species, but the horse and all of its nonnative friends are out of the barn already. The government can regulate people, but nature is not so receptive. It is a challenging issue, to say the least.

Louisiana has tried two approaches worth noting with respect to nutria: paying bounties and encouraging a new cuisine. With federal funding through the Coastal Wetlands Planning, Protection, and Restoration Act, the Louisiana Department of Wildlife and Fisheries pays hunters and trappers $5 a head (or in reality, a tail). In the 2008–2009 season, 262 participants received $1,670,190 for killing 334,038 nutria by trap, rifle, or shotgun (Weibe and Mouton, 2009). The program has contributed to a reduction in areas damaged by nutria (from about 80,000 to 20,000 acres), while also providing much opportunity for shooting practice.

Nutria Chili

Recipe by: Chef Enola Prudhomme

3 tablespoons vegetable oil
2 pounds nutria ground meat
1 tablespoon + 1 teaspoon salt
1 teaspoon red pepper
1 tablespoon + 1 teaspoon chili powder
1 cup diced onion
1 cup diced green bell pepper
1 cup diced red bell pepper
1 cup tomato paste
4 cups beef stock (or water)
1 can red kidney beans (opt.)

In a heavy 5-quart pot on high heat, add oil and heat until very hot. Add nutria meat, and cook and stir 10 minutes. Add salt, red pepper, chili powder, onion and both bell peppers. Cook and stir 15 minutes. Add tomato paste and 4 cups stock. Cook 30 minutes; reduce heat to medium. Add red kidney beans; cook an additional 10 minutes. Serve hot!

FIGURE 6-4. Innovative Cajun cuisine. (Source: Louisiana Department of Wildlife and Fisheries.)

Another "market"-based approach was to encourage a culture of eating nutria. Chef Enola Prudhomme was recruited in the effort, and nutria recipes are posted on the Web site of the Louisiana Department of Wildlife and Fisheries (figure 6-4). (If you have a broader palate, consult the *Invasive Species Cookbook*, published by the Bradford Street Press.)

Reportedly, nutria tastes more like rabbit than chicken. Some have even suggested renaming it "bayou rabbit" for marketing purposes. Although it is a plentiful source of low-fat protein, the fact that it is a rodent (and frequent roadkill) has tended to dampen the public's demand.

Net gains on agricultural lands

First the good news: Not only have wetland losses due to agricultural activities declined, but we have actually seen a net gain in wetland acreage in farming areas. The disincentive of Swampbuster and the affirmative payments of the Wetlands Reserve Program appear to be working. There is also an argument that unfavorable economic conditions for farming have contributed to the decline. Nevertheless, from the mid-1970s to the mid-1980s, agriculture was responsible for an estimated 137,500 acres of wetland losses annually (Dahl and Johnson, 1991). The trend has been reversed. During the period 1998–2004, the FWS reports that agriculture was responsible for a net gain of 70,700 acres, or more than 11,500 acres per year (Dahl, 2006).

An example from New Jersey can help illuminate these dry statistics. Although New Jersey has long been derided as a pollution haven (and a trip on the New Jersey Turnpike does nothing to dispel the reputation), it remains, as its license plates attest, the Garden State. New Jersey has more than 733,000 acres of farmland (with 488,000 acres of cropland), and it is the nation's third-highest cranberry producer, behind Wisconsin and Massachusetts. It is also the home of the Franklin Parker Reserve, the largest Wetlands Reserve Program project in the Northeast.

The Franklin Parker Preserve is a 9,400-acre freshwater wetland complex located in the middle of New Jersey's Pine Barrens. Portions of the site had been used for cranberry operations, which required the manipulation of the water regime through dikes and canals. Although still an aquatic environment, the cranberry bogs were a highly modified environment and not functioning as native wetlands. In 2003, a land trust organization, New Jersey Conservation Foundation, purchased the site. Two years later, it enrolled 2,200 acres into the Wetlands Reserve Program. The Natural Resources Conservation Service paid $4.4 million for a conservation easement and will also contribute $1 million to restoring the site. The cost comes out to less than $2,500 per acre.

In several respects, the Franklin Parker Preserve highlights certain factors that are necessary for the long-term success of many restoration sites. It is large, it is located adjacent to other protected lands—250,000 acres of state-owned property—and it is owned by an entity that is devoted to its long-term stewardship.

The performance of Clean Water Act mitigation sites offers a stark and a more depressing contrast.

Paper gains, real losses: The failure of permittee-responsible compensatory mitigation

For many years, the failure of compensatory mitigation was the Clean Water Act section 404 program's dirty little secret (Gardner, 1996). It was similar to the old Soviet joke: we pretend to work, and they pretend to pay us. Permittees would pretend to provide functioning wetland mitigation, and the Corps would pretend that it offset wetland impacts from permitted activities.

State and federal studies in the early 1990s suggested that the scope of the failure of compensatory mitigation projects was massive. In 1990 the Florida state legislature requested the Florida Department of Environmental Regulation (DER) to review the performance of compensatory mitigation projects within the state. Examining sixty-three permits that required wetland creation to offset development impacts, the DER found a staggeringly high rate of noncompliance (Florida DER, 1991). A mere four permittees—only 6.3 percent—had complied with their mitigation requirements. About a third of the permittees had not even bothered to attempt their mitigation projects, even though the damage to wetlands had already occurred.

Across the country in the state of Washington, a 1994 federal study of seventeen restored and created mitigation sites yielded similar results. The EPA and the FWS noted that two developers had taken no action to begin the required compensatory mitigation. On the remaining sites, the agencies discovered that eleven—nearly two-thirds—failed to provide an acceptable level of ecological functions. Only four mitigation projects were functioning well.

In 1999, the EPA sought an objective and comprehensive review of wetland compensatory mitigation success and failure from the National Academy of Sciences. Although it was created by a congressional charter in 1863, the National Academy of Sciences is not an agency; it is a private, nonprofit organization that advises the federal government on scientific and technical matters. The primary operating arm of the National Academy of Sciences is the National Research Council (NRC). Pursuant to the EPA's request (and a cooperative agreement that provided funding), the NRC formed a thirteen-member interdisciplinary committee (formally known as the Committee on Mitigating Wetland Losses) to evaluate mitigation practices under the Clean Water Act. The NRC committee was led by Dr. Joy Zedler, a botany professor and Aldo Leopold Chair of Restoration Ecology at the University of Wisconsin, and included other wetland ecologists, soil scientists, hydrologists, environmental planners and consultants, an economist, and even a lawyer.

After two years of examining scientific literature, meeting with experts and stakeholders, and conducting site visits, the NRC committee issued a report that confirmed the concerns about compensatory mitigation. The

NRC committee found that despite some progress in the past twenty years, the "goal of no net loss of wetlands is not being met for wetland functions by the [Clean Water Act section 404] mitigation program." On paper, there appeared to be no net loss—even a significant net gain of wetland acreage. From 1993 to 2000, the Corps permitted approximately 24,000 acres of wetlands per year to be filled. As a condition of those permits, the Corps required about 42,000 acres of compensatory mitigation each year. This works out to a 1:1.8 ratio; that is, for every acre of wetland filled, the Corps required 1.8 acres in compensatory mitigation. But looking behind the numbers, the NRC committee found it was much too early and misleading to declare victory.

First, replacement of lost acreage does not necessarily equal replication of lost functions and ecosystem services. Wetlands are not fungible like money (or even a ton of carbon dioxide). Different wetlands provide different functions, and the value of their services depends greatly on their position in the landscape. Consider a 10-acre riverine wetland that provides flood storage capacity when the river overflows its banks, improves the river's water quality by filtering land-based pollution, and offers habitat for migratory waterfowl. The value of these ecosystem services (to humans) is contingent on a number of factors, including what property is protected from flooding, how people use the river (drinking source, fishing, recreation), and whether the bird species is of interest to birders or hunters. If the Corps permits the riverine wetland to be filled for a housing development, it might require 10, 15, or maybe 20 acres to be restored. But whether the restoration project truly offsets the environmental impacts of the development activity depends on the functions and services lost due to the development, the functions and services gained from the mitigation project, and the position of the wetlands (filled and mitigation) on the landscape. Such a determination is obviously very complex, and the Corps traditionally has not tracked wetland functions or services lost and gained. Instead, it has used acreage as a rough surrogate. Thus, it is not possible to find that no net loss for wetland function or ecosystem services has been met.

Even relying on the questionable metric of acreage, however, we cannot declare that the Clean Water Act section 404 program has achieved no net loss of wetlands. Such a conclusion assumes that the mitigation projects, or at least a majority of them, were successful. Yet while wetland losses (in terms of acreage, functions, and ecosystem services) were real, mitigation success was ephemeral.

The committee found that mitigation failures were common for various reasons. Sometimes the permittee would not even commence the promised restoration, enhancement, or creation project. Sometimes the permittee would try, but the mitigation project would fail, perhaps due to improper elevation or siting that led to insufficient water or too much water. It is relatively easy to create a pond; it is much more difficult to create a self-sustaining wetland. A successful wetland mitigation project requires a commitment of time and money, as well as technical expertise. Permittees were unwilling or unable to provide the necessary resources, and as a result compensatory mitigation failure was rampant.

Even when a mitigation project was initially successful legally (meeting any performance standards specified in the permit) and ecologically (replacing lost functions), long-term prospects were questionable. Once the Corps approved the mitigation project as completed, it was assumed that the new wetland would be fine on its own. Once the Corps signed off, that ended the permittee's responsibility for the site. It was presumed that nature would take over. But the NRC committee cautioned that the "presumption that once mitigation sites meet their permit criteria they will be self-sustaining in the absence of any management or care is flawed." A mitigation site may need to be protected from trespassers, hydrological adjustments may need to be made, and prescribed burns may need to be periodically conducted. Furthermore, a mitigation site might need to be protected from invasive species. But for almost all permits, once the Corps gave its stamp of approval on the mitigation site, no one was responsible for its long-term stewardship.

An agency such as the Corps has enforcement powers, and one might assume that when confronted with mitigation failures, the Corps would take action. Indeed, it has a number of weapons in its arsenal. If a permittee fails to comply with the conditions of its permit, the Corps has the legal authority to revoke the permit or bring an administrative or judicial enforcement action to force the permittee to complete the promised mitigation. The NRC committee found, however, that monitoring of mitigation sites and enforcement of mitigation conditions were not a priority within the Corps.

Corps headquarters had issued "standard operating procedures" (SOPs) for Corps districts. The SOPs divided regulatory activities into two categories—those above the line and those below the line. "Above the line" activities were those items that Corps regulators needed to accomplish; they were the priorities. The "below the line" activities were deemphasized, and Corps regulators were to turn to these items only after having accomplished the above the line activities. The priority item above the line was issuing permits. Below the line were activities such as compliance inspections for mitigation and multiple visits to a mitigation site.

With insufficiently clear performance standards, sparse agency monitoring and inspections, and almost nonexistent enforcement of mitigation conditions, the Clean Water Act section 404 program was not achieving no net loss for either wetland functions or acreage. The NRC committee did, however, provide recommendations on how to improve the situation. The Corps should increase its reliance on the first step of the sequence; in particular, impacts to wetlands that are difficult to restore, such as bogs and fens, should be avoided. The committee also suggested that the Corps adopt a landscape approach to permit-decision making, considering the whole watershed with respect to filling activities and mitigation projects. The committee also recognized that alternatives to permittee-responsible mitigation, such as wetland mitigation banks or in-lieu fee programs, offered some advantages. Mitigation banks and in-lieu fees will be taken up in the next chapters. While they represent an improvement over the traditional approach of permittee-responsible mitigation, they are not without controversies or critics.

No net loss: Mission accomplished?

In light of the abysmal failures of compensatory mitigation in the Clean Water Act section 404, how could departing Secretary of Interior Gale Norton declare that we achieved a net gain in wetlands during 1998–2004? On the one hand, it was an accurate characterization of a FWS report that found that the United States had an annual net gain of 32,000 acres in wetland *area* over that time period. But the report contains many caveats, and the net gain in this case depended in part on what is considered a wetland. As the agency candidly acknowledged, its methodology included ponds (such as golf course water hazards) as wetlands—as had the earlier status and trends reports. Although it was the inclusion of golf course water hazards that brought ridicule, the larger point is that the ponds (whether on a golf course or not) do not provide the same functions as wetlands and consequently do not provide the same level of ecosystem services. No net loss in wetland *area* does not necessarily equate to no net loss of wetland *functions*. Indeed, the FWS report stated as much. So a declaration of achieving a net gain may be technically accurate, but it is also incomplete and misleading.

A 2008 study provided a more nuanced picture. During 1998–2004, wetlands in the coastal watersheds of the eastern United States were being diminished at a rate of 59,000 acres per year (Stedman and Dahl, 2008). Such a figure underscores the hollowness of focusing exclusively on wetland area. The EPA is, however, embarking on a National Wetland Condition Assessment, which would be a more qualitative examination of the nation's wetlands. A report is due out in 2013.

In the interim, federal and state agencies will continue to ponder how best to create incentives to deal with invasive species, to encourage farmers to be good environmental stewards of the land, and to offset wetland impacts from permitted activities. We now turn to the vexing question of how to improve the effectiveness of compensatory mitigation and one particular proposed tool: wetland mitigation banking.

Chapter 7

Wetland Mitigation Banking: Banking on Entrepreneurs

[F]or what is land but the profits thereof[?]

—*Sir Edward Coke, 1628*

Sir Edward Coke, a seventeenth-century English jurist, suggested that the value of real property is intrinsically tied to the ability to exploit its resources: to farm the land, to harvest its trees, to mine its ore. Almost four centuries later, this view continues to hold sway. Indeed, property (wetlands or otherwise) left in its natural state is often considered to be economically idle. Is it possible, however, for undeveloped land to yield a profit? May an owner benefit economically by deciding not to develop land or by actively restoring it to its natural condition? Until recently, the answer was almost always "no." But the rapidly growing business of wetland mitigation banking has changed the calculus. If properly implemented, mitigation banking can offer economic benefits to private landowners and ecological benefits to the public.

Wetland mitigation banking remains controversial, but has the potential to contribute to the goal of no net loss of wetlands. Its proponents describe it as the proverbial "win-win" situation. Its critics suggest it is a modern-day version of selling Florida swampland to unsuspecting investors, only this time the public will bear the costs. Regardless of your view, mitigation banking provides an intriguing administrative law case study: it is a "market-based" approach to environmental protection in

111

which the government controls both the supply side and the demand side of the equation. It is also an industry that was initially built on guidance.

What is wetland mitigation banking?

As explained in the last chapter, the Corps typically grants a section 404 permit on the condition that the permittee will offset its adverse wetland impacts through compensatory mitigation (such as a restoration project). Traditionally, permittees performed the compensatory mitigation project concurrent with or after the development project, either by doing the mitigation work itself or by hiring an environmental consultant or engineering firm. And, as noted, this permittee-responsible mitigation did not fare well. Permittees had little incentive to ensure that the mitigation resulted in a functioning wetland. They had already completed or were in the process of completing their development project, and the Corps generally showed little interest in mitigation inspections and enforcement actions. Mitigation banking seemed to offer an alternative: if the mitigation project was done in advance of impacts, rather than after the fact, there should be a greater likelihood of success.

Thus in its first iteration, mitigation banking was defined as advance mitigation. Among the earliest mitigation bankers were state departments of transportation (DOTs) (Environmental Law Institute [hereafter ELI], 2002). The Montana DOT, for example, knew that it would be constructing miles and miles of linear road projects and that it would be impossible to eliminate all wetland impacts. Accordingly, Montana DOT was also aware that it would need to offset those wetland impacts. Instead of waiting to do the mitigation projects after highway projects, it entered into a memorandum of agreement (MOA) with the Corps. Montana DOT would do a restoration project before the highway project; when the restoration project met certain performance standards, the Corps would award the Montana DOT "wetland credits." These credits would then be "banked" for future use. The mitigation bank was, in effect, a savings account. The Montana DOT would withdraw the banked credit when it needed to provide compensatory mitigation for a later highway project. This is sometimes called a single-user bank, since the bank is managed for the benefit of one entity.

Single-user banks were not limited to state DOTs.[1] The Walt Disney Company entered into an arrangement in the early 1990s that, while not called a mitigation bank at the time, certainly was a forerunner and model of advance mitigation (Gatewood, 1995). Disney was faced with long-range planning challenges. It had a comprehensive build-out plan for Disney World, which would affect 448 acres of wetlands on-site. Most Corps section 404 permits, however, are valid for only five years; if a permittee has not started its project within that time, it might need to start the permit process over. Five years was too short a time horizon for Disney's strategic plan. In discussions with the Corps, Disney requested a twenty-year permit and suggested an alternative to on-site mitigation: Disney would acquire Walker Ranch, an 8,500-acre site that was under the threat of development, and restore wetlands there. In terms of environmental benefits, the Walker Ranch option appeared to be better than the on-site creation option. First, more acres would be restored (and restoration was favored over creation), and many acres would be restored before development impacts occurred. Second, Walker Ranch was a key parcel that provided ecological connectivity between public lands to its east and west. Third, Disney ensured the long-term management of the site by enlisting The Nature Conservancy to serve as its steward.

The Walker Ranch option also benefitted Disney economically. To be sure, the off-site mitigation, which would include 752 acres of restoration, 1,800 acres of enhancement, and 1,350 acres of preservation, was more expensive than the on-site creation of 600 acres: the off-site mitigation was estimated to cost $40–$45 million versus $38.5 million for the on-site work. But by moving the mitigation off-site, Disney was then able to retain the use of 600 acres of its property that otherwise would have been created wetlands. If that property is assigned a conservative value of $30,000 per acre, then the off-site option at Walker Ranch made financial sense (table 7-1).

In theory, wetland mitigation banking offered several ecological advantages over the traditional approach to compensatory mitigation (Gardner, 1996). First, because it was advance mitigation (done before development), mitigation banking reduced temporal losses (the time period between wetland impacts from a development project and wetland gains from a successful mitigation project). Second, mitigation banking offered a greater

TABLE 7-1. On-site versus off-site mitigation expenses for Disney.

- On-site mitigation
  - $38.5M + 600 acres of Disney property
  - 600 acres @ $30,000 = $18M
- Off-site mitigation
  - $40–$45M

---

likelihood that projects would be completed successfully. If performance standards were not met, then no credits were generated; the mitigation banker would have to try again. Contrast this with the traditional approach, where the permittee promised to provide compensatory mitigation, but frequently failed to do so, with few repercussions. Mitigation banking also was more likely to result in larger wetland sites, which could be more sustainable than the so-called postage stamp mitigation. Moreover, because of economies of scale, mitigation banking was thought to provide more cost-effective (cheaper) means of compensatory mitigation.

Entrepreneurs began to look at the mitigation banking model to see if there was a role for the private sector in the development and sale of wetland credits. Could a private company restore a degraded wetland and sell the resulting credits to a permittee who needed to satisfy its compensatory mitigation obligations? One of the initial barriers was the lack of a clear legal framework that authorized wetland mitigation banking.

The legal status of mitigation banking (the early years)

As noted previously, the Clean Water Act does not use the term "mitigation." Thus, it should be no surprise that it also contains no reference to mitigation banking. The concept of producing and trading wetland credits was foreign to the 92nd Congress in 1972. But mitigation banking was also absent in agency regulations. It was not mentioned in either the EPA's section 404(b)(1) Guidelines in 1980 or the Corps' regulations. Mitigation banking made its first, brief appearance in a formal agency document in guidance. The 1990 Mitigation MOA between the Corps and the EPA announced that mitigation banking "may be an acceptable form of compensatory mitigation." The MOA suggested that "an environmentally successful bank" could be used as long as the Corps and the EPA approved. Further guidance on the subject was promised.

Three years later the Corps and the EPA issued the additional guidance, in the form of a four-page memorandum to the field. The August 1993 memo characterized itself as interim guidance, pending the results of mitigation banking studies by the Corps' Institute for Water Resources. The memo reinforced the need to follow the mandatory sequence of avoid-minimize-compensate. Mitigation banking was not to be an excuse to avoid the avoidance requirement. The memo's definition of mitigation banking emphasized that it was advance mitigation:

restoration, creation, enhancement, and, in exceptional circumstances, preservation of wetlands or other aquatic habitats expressly for the purpose of providing compensatory mitigation *in advance of discharges into wetlands* permitted under the Section 404 regulatory program. (emphasis added)

Nevertheless, the memo also recognized that the agencies could authorize the release of credits before the bank was fully functioning, to ensure that the bank was financially viable. In such cases, it would be appropriate to impose a higher ratio (e.g., 2 acres from the bank to offset 1 acre of impacts). As we will see, the issue of early release of credits came under heavy criticism from environmental groups and eventually led to a revised (and more realistic) definition of mitigation banking.

Significantly, the 1993 memo added flesh to the bone by listing what elements should be included in an agreement between the bank sponsor and the agencies. Among other items, the agreement needed to establish the baseline ecological conditions of the mitigation site; a methodology for evaluating progress; accounting procedures for tracking credits and debits; the geographic service area (i.e., the potential customer base—the area in which credits could be used); monitoring requirements; and provisions for perpetual stewardship of the site. In addition, the memo also recognized that entrepreneurial banking could be permissible and wetland credits could be bought and sold. The memo emphasized, however, that if a permittee bought a mitigation bank credit, the permittee retained ultimate responsibility for meeting its compensatory mitigation obligations. The sale of the credit was not (yet) expressly a transfer of liability.

Pembroke Pines: The first sale of credits from an entrepreneurial mitigation bank

Although the 1993 interim guidance provided a framework for mitigation banking, it remained a risky endeavor in which to invest. A mitigation banker needed to spend money to prepare environmental studies, acquire a mitigation bank site, navigate through a new and uncertain regulatory process (at local, state, and federal levels), and begin the biological improvements, such as removing invasive species and planting native vegetation. If the mitigation bank then met certain performance standards, credits could be released for sale. If the performance standards were not met, then the banker would have few if any credits to sell. Accordingly, it could take years for a mitigation banker to make a return off its investment.

One of the earliest mitigation bank companies, Florida Wetlandsbank, Inc., whose officials included a former developer, realized that it would take years to get an entrepreneurial bank approved in south Florida (Gardner, 2003). They knew that the carrying costs of land (e.g., property taxes, mortgage payments) in south Florida would be prohibitively expensive. To minimize the costs and thus the risks, the company entered into an arrangement with the City of Pembroke Pines. The company agreed to restore 445 acres of city-owned wetlands that were overrun and choked with melaleuca. The company removed the invasives, regraded the land, and planted native wetland vegetation. The vegetation survived (after some replanting), the marsh thrived, and wildlife returned (more than 120 species have been documented on the site). Because the mitigation site met its performance standards, the agencies okayed the release of credits for sale.[2]

When the company sold a credit, part of the proceeds went to the city. A portion of the city's share was dedicated to the long-term maintenance of the site, and part could be used by the city for other purposes. (It was reported that the city received $3.5 million in licensing fees and an additional $500,000 was set aside in a trust fund for maintenance of the site.) The first credit was sold in January 1994. The mitigation bank sold its final credits in 1999, and the city now manages the area as an environmental park (with boardwalks).

The Pembroke Pines Bank was a vast improvement over traditional permittee-responsible mitigation. The project met its performance standards, and at the end of the day it had a steward (the city) with a trust fund to care for the site. It appeared to be a net gain for everyone involved. The city had a new park, and the company was rewarded financially for its risks. It was easier for the agencies, which now could inspect a single, large mitigation site instead of trying to track down dozens of smaller sites. Even the permittees who purchased the credits were happy. For a fixed price, they were able to satisfy their compensatory mitigation obligations. State and federal agencies had for the first time approved shifting the responsibility for the compensatory mitigation from permittees to a mitigation banker (a practice that would soon be expressly authorized in federal guidance).

On one level, it appears that the mitigation banker is selling a wetland credit. In reality, the mitigation banker is selling a release or transfer of liability. The permittee is purchasing peace of mind: legal and financial certainty. If the mitigation bank site has problems—if it is an unusually dry (or wet) year and the plants die; if feral hogs or nutria invade the site and cause damage—the permittee does not need to worry. That is now the mitigation banker's problem.

As the Pembroke Pines Bank demonstrates, there may also be a second transfer. Once the credits are sold, once the site has met its performance standards, a mitigation banker may transfer the responsibility to care for the site to another party, such as the government or a land trust or conservancy organization (NRC, 2001). Ideally, as was the case with the Pembroke Pines Bank, the stewardship organization will receive ample funds to manage the site.

But if the Pembroke Pines Bank offered a model on how to improve compensatory mitigation, not many entrepreneurs followed its lead. In the early 1990s, almost all mitigation banks were publicly owned banks, such as those run by state DOTs or port authorities (ELI, 1993). If the federal agencies wanted to increase private investment in wetland mitigation banking, they were going to need to provide a firmer legal foundation for the enterprise.

The 1995 mitigation banking guidance

Thus far, the rules governing mitigation banking were only in guidance documents. While the guidance documents were available to the public, they did not go through any public notice-and-comment process in their development. The next major step that the agencies took was to issue even more detailed guidance, only this time they did so through a notice-and-comment process.

In March 1995, the Corps and the EPA, along with the FWS, NRCS, and National Marine Fisheries Service (NMFS), published in the *Federal Register* a proposed rule on the establishment and operation of wetland mitigation banks. Although the proposed rule was interpretive and a statement of policy—and not a binding regulation—the agencies nevertheless sought public input. This input was not legally required, but the agencies decided that because of the controversial nature of the proposal, the final guidance document would benefit from a public vetting. After receiving more than 130 comments, the agencies published a final version of the guidance in November 1995.

Building on the 1993 interim guidance, the 1995 guidance set forth a more detailed process for establishing a mitigation bank. First, a bank sponsor, the entity that would be responsible for the mitigation site, was encouraged to meet with the various agencies in a pre-application process. This pre-application process (which is common for large-scale development projects as well) should serve to identify potential areas of agreement and disagreement. The bank sponsor should then submit a prospectus to the Corps, which will be the basis of the mitigation bank instrument. The mitigation bank instrument is a comprehensive document that governs how the bank is established, how it generates credits, how and where the credits may be sold, what financial assurances are required, and how the site will be monitored and maintained. In a nutshell, the mitigation banking instrument sets the rules for a particular mitigation bank.

The 1995 guidance established an interagency mitigation banking review team (known as the MBRT), which approved mitigation banking instruments and oversaw the operation of mitigation banks. The composition of the MBRT varied from Corps district to district, but typically included representatives from the Corps, the EPA, the FWS, and perhaps NMFS and state and local agencies. The Corps was designated the lead agency in the MBRT, but the guidance envisioned the process playing out on a consensus-based approach. While in theory having more wetland and wildlife experts in the room leads to a better end product, a practical difficulty turned out to be getting all the representatives in the same room at the same time. Mitigation bankers have querulously noted the long delays in mitigation banking reviews.[3]

The 1995 guidance also elaborated on a host of policy considerations related to mitigation banks. It again emphasized the need to follow the sequence of avoid-minimize-compensate. To bolster the financial security of banks, the guidance recognized that a certain percentage of credits could be released early, allowing "limited debiting of a percentage of the total credits projected for the bank at maturity." Banks could be established on private and public lands, although credit generated from the latter would be restricted to benefits the bank provided beyond current or planned public programs. Restoration was the preferred form of compensatory mitigation over enhancement and creation. Preserving wetlands could be used to augment a bank's credits from other forms of compensatory mitigation, but preservation-only banks were permissible only in "exceptional circumstances." The guidance also clarified that when a mitigation bank sold a credit, it then assumed the responsibility for the permittee's compensatory mitigation obligations.

Thus the 1995 guidance provided more detailed rules. It crafted a complex procedure for bank approval and operation. Importantly, it allowed for a mitigation bank to sell a release or transfer of liability. You might think then that this is a sufficient legal framework for entrepreneurial banks to flourish. But before you open your checkbook to invest, note this nugget from the guidance:

> The policies set out in this document are not final agency action, but are intended solely as guidance. The guidance is not intended, nor can it be relied upon, to create any rights enforceable by any party in litigation with the United States. This guidance does not establish or affect legal rights or obligations, establish a binding norm on any party and it is not finally determinative of the issues addressed. Any regulatory decisions made by the agencies in any particular matter addressed by this guidance will be made by applying the governing law and regulations to the relevant facts.

In other words, guidance is simply that: guidance. It is not "final agency action," so it is not ripe for a judicial challenge. It does not create any legal rights or obligations, so rely on it at your own peril. And as a guidance document, it can be revoked at any time. Have a nice day.

To be sure, the 1995 guidance did provide sufficient comfort to some investors, and the number of entrepreneurial banks began to rise. But an even bigger boost to the fledgling industry came from Congress.

Congress provides a market (and ratifies the guidance).

In 1998, Congress waded into the mitigation debate in the Transportation Equity Act for the 21st Century, known as TEA-21. TEA-21 authorized more than $200 billion to be spent to improve highways, roads, bridges, and other transportation infrastructure. Up to 20 percent of the funds for a particular transportation project could be devoted to offsetting the environmental impacts, including wetland mitigation. Most critical, from the mitigation banking industry's perspective, was section 1108, in which Congress declared that for federally funded transportation projects

> preference shall be given, to the maximum extent practicable, to the use of the mitigation bank if the bank contains sufficient available credits to offset the impact and the bank is approved in accordance with the Federal Guidance for the Establishment, Use and Operation of Mitigation Banks (60 Fed. Reg. 58605 [Nov. 28, 1995]) or other applicable Federal law (including regulations).

Through this preference, Congress had created a ready-made market for credits from mitigation banks.

Consider, as well, the unusual administrative law aspect of this provision. Ordinarily, Congress enacts a statute (e.g., the Clean Water Act). An agency then promulgates a regulation to implement the statute (e.g., EPA's section 404(b)(1) guidelines). The agency then might issue a guidance document that further clarifies the statute and regulation (e.g., the 1995 mitigation banking guidance) and that is the end of the process. But here Congress referenced and in effect ratified the 1995 mitigation banking guidance. It is extraordinarily rare for Congress to provide its imprimatur in a statute for a guidance document.[4] Although the Clean Water Act and (at that point) Corps and EPA regulations said nothing about wetland mitigation banking, TEA-21 boosted the legitimacy of the mitigation banking concept.

Fueled in part by congressional support (and a statutorily created customer base), the number of mitigation banks increased dramatically, especially privately operated entrepreneurial banks. In 1993, the Environmental Law Institute reported that 46 mitigation banks had been established, and only one was an entrepreneurial bank. Almost ten years later, there were 219 mitigation banks approved, and more than 60 percent were entrepreneurial (ELI, 2002). By December 2005, 405 banks had been approved, 72 percent of which were entrepreneurial banks (Wilkinson and Thompson, 2006). And the numbers continue to increase. At the National Mitigation and Ecosystem Banking Conference in 2010 in Austin, Texas, it was reported that there were now nearly a thousand approved mitigation banks, with another 500 banks in some stage of development.

One reason for the increase is that mitigation banking is viewed as a "license for printing money," as one banker colorfully described it (Pittman and Waite, 2009).

How much can I sell a wetland credit for?

That is almost always the first question that someone who is thinking about entering into the mitigation business asks. They see reports that credits are selling in some parts of the country for more than $100,000 per acre, and they figure that it is easy to get big returns. But focusing on the (reported) sales price of a credit overlooks the initial investment needed and the financial risks associated with mitigation banks.

Let's work through the numbers in a hypothetical mitigation banking pro forma (which is loosely based on presentations given by a mitigation banker). Assume that the bank site is a 125-acre parcel with a mixture of de-graded freshwater and tidal wetlands, which can be purchased for $30,000 per acre—so the initial land costs are $3.75 million. Also assume that after restoration, the site will contain 50 acres of freshwater wetlands and 50 acres of tidal wetlands, with 25 acres of upland buffer. To keep things relatively simple, let us also calculate credits on an acre basis. With 50 acres of freshwa-ter wetlands and 50 acres of tidal wetlands, you might think that the bank will yield 100 credits. This is understandable, but most likely wrong.

Mitigation banks generate credits based on the ecological improve-ments they provide. To determine the credit generated, you need to know the baseline condition of the site. A heavily degraded site (such as an agri-cultural area that once was a wetland, but after being ditched and drained is now a cornfield) has more potential to produce credits than a relatively pristine site. What's critical is the *delta*, the difference between current site conditions and future site conditions (assuming that the restoration project is successful).

Turning back to our hypothetical site, let us say that the freshwater wet-lands are only partially degraded, infested with an invasive species, but still providing some ecosystem services. The regulatory agency decides to give you only 0.5 credits per acre for the freshwater wetlands, for a total of 25 credits. But the good news (if you are a banker) is that the tidal wetlands are in terrible shape, and the agency is willing to give you 1 credit per acre, for a total of 50 credits. You will receive no wetland credits for the 25 acres of upland, however. Thus, your 125-acre mitigation bank has the potential to yield 75 credits.

Seventy-five credits sold at $100,000 each would result in $7.5 mil-lion. If the land was $3.75 million, it would appear that you have made 100 percent return on your investment! There are, of course, other costs and considerations that may dampen your enthusiasm. Funds need to be set aside for the endowment of a trust fund for long-term management of the site ($750,000), and there are expenses associated with marketing and sell-ing the credits ($50,000). You must also factor in the cost of the restoration project, removing invasive species, plugging ditches, regrading the land, and planting native vegetation ($750,000). And there are general and ad-ministrative expenses, such as insurance premiums and environmental con-sultants for the permitting process—and attorney fees ($400,000). Now your return is down to $1.8 million, which still looks good (table 7-2).

Recall, however, that mitigation banking is supposed to be "advance" mitigation, and you can only sell credits as you meet certain performance standards that demonstrate that the bank site is progressing toward its

TABLE 7-2. A mitigation banking pro forma.

Credit Calculations		Cost Calculations	
Freshwater Wetlands (50 acres @ 0.5 credit/acre)	25 credits	Land Acquisition (125 acres @ $30,000)	$3,750,000
Tidal Wetlands (50 acres @ 1.0 credit/acre)	50 credits	Long-term Endowment	750,000
		Sales and Marketing	50,000
Upland Buffer (25 acres @ 0 credit/acre)	0 credits	Construction/Restoration Work	750,000
Total Credits	75 credits		
		General and Administrative	$ 400,000
Cost per Credit	$ 100,000		
Total Sales	**$7,500,000**	**Total Costs**	**$5,700,000**

desired state. The Corps will release a certain number of credits for you to sell after your mitigation banking instrument is approved, the property is acquired, and a conservation easement is placed upon it. Yet the rest of the credits will likely be released in phases over a five-year period—if the restoration project satisfies its performance standards. If the restoration project is not successful, you will have no more credits to sell.

An additional element of uncertainty is that the market for credits is created by the government, and government policies change from time to time. When the U.S. Supreme Court invalidated the Migratory Bird Rule in *Solid Waste Agency of Northern Cook County*, it was not just environmentalists who were sad. The decision removed Clean Water Act jurisdiction from certain isolated wetlands, and with it the need to provide mitigation for filling those wetlands (for federal purposes). Some mitigation bankers saw the demand for their credits immediately drop nearly in half (Robertson, 2006).[5]

Of course, there are ways in which a mitigation banker can reduce costs. As we saw with the Pembroke Pines Mitigation Bank, the banker may enter into an agreement with a landowner, thereby avoiding land acquisition costs. (Remove the $3.75 million land costs from the equation, and the potential return looks very different.) Income will be reduced, however, as the landowner will want to receive a share of the credit sales. Mitigation bankers have entered into such relationships with farmers and ranchers who see housing developments and condominiums sprout on their neighbors' lands. The farmers and ranchers may want to retire or they may want some added income, but they may not want to sell out to developers. A mitigation banker offers an alternative. One of the more unusual relationships is the Monastery Mitigation Bank in Conyers, Georgia. The Trappist monks of the Monastery of the Holy Ghost agreed to allow their land to be part of a mitigation bank. In return, they receive a percentage of the credit sales, which augments the income they generate from selling jams and jellies.

There are several other strategies that mitigation bankers can use to reduce costs and risks. When negotiating a mitigation banking instrument, for example, a banker will seek to maximize the number of credits the bank can provide, accelerate the credit release schedule, and expand the geographic service area where the credits can be used. More credits, more readily available, and more customers are good for business; however, they can be in tension with the environmental goals of agencies. The agencies do not want to give credit unless there are demonstrated ecological gains, and they generally want these gains to be close to the impact sites. Early credit releases can increase the risk of a bank failure (i.e., a bank sells its credits but fails to follow through with the promised mitigation), and large service areas can contribute to the migration of wetlands and their ecosystem services from urban and suburban areas to more rural areas where land is typically cheaper (Ruhl and Salzman, 2006). The objectives of the mitigation banker and the regulatory agency are not always perfectly aligned.

The good, the bad, and the ugly

The Corps and the EPA (and Congress and the National Research Council) recognize that mitigation banking is an improvement over traditional permittee-responsible mitigation. While some mitigation banks have been financial and ecological successes, others banks have not. Lessons can be drawn from both categories. Let us consider several examples.

Panther Island Mitigation Bank (Florida)

As in real estate, the three most important things about a mitigation bank are location, location, and location.[6] Panther Island Mitigation Bank (PIMB) was established adjacent to the Audubon Society's Corkscrew Swamp Sanctuary in Naples, Florida. Corkscrew Swamp Sanctuary was, at the time, approximately 10,200 acres and was (and continues to be) home to the world's largest remaining stand of virgin bald cypress. The sanctuary's 500-year-old trees provide critical nesting areas for the endangered wood stork. After an eighteen-month MBRT review process, PIMB received a federal mitigation banking instrument that authorized up to 934 mitigation credits for restoration, enhancement, and preservation of the 2,778-acre site. Cattle and invasive species were removed, and marshes were restored.

The mitigation bank site, with its marshes, sloughs, and cypress domes, has been turned over to the Audubon Society, which has incorporated it into its sanctuary. Furthermore, PIMB provided the Audubon Society funding to manage the newly acquired property. The end result should serve as a model for all compensatory mitigation efforts: a restored and preserved site that is now owned by a dedicated steward that has sufficient funding for long-term management responsibilities.

Mud Slough (Oregon)

Mitigation banks do not need to be thousands of acres in size, as Mud Slough, a 56-acre bank located on former agricultural lands, demonstrates. The landowners had enrolled other parcels in the Wetlands Reserve Program and decided to try their hand at mitigation banking. The ecological results were very positive: the Audubon Society of Portland lists the site as an important bird area, and the Oregon State Land Board bestowed the 2003 Wetland Project Award upon it. The landowners are vigilant in managing the site and eradicating invasive species such as reed canary grass. There is a blip on the horizon, however. While the mitigation banking instrument required that a deed restriction be placed on the mitigation bank site, it did not require a long-term maintenance fund. Thus, there is no requirement that an escrow account or trust fund be established for the management the property. That is not an immediate problem as the current landowners are devoted to the land. But what will happen when the land is transferred? While the deed restriction may prohibit active destruction of the restored wetlands, it does not require the property owner to actively manage them.

South Carolina Department of Transportation Black River Mitigation Bank (South Carolina)

This early mitigation DOT bank, approved in 1997, represents one extreme of the continuum of geographic service area. This 1,709-acre site in

Clarendon County was a bottomland hardwood swamp ditched, drained, and logged. The agreement with the Corps and other agencies was that South Carolina DOT could use the credits generated from restoration to offset unavoidable wetland impacts related to highway construction projects. From an environmental perspective, the good news was that the mitigation to impact ratio ranged from 3:1 to 4:1 (i.e., 3 to 4 acres of mitigation would be required for each acre of impact). But the bad news— shocking, in fact—was that the service area was the entire state of South Carolina and its five major river basins. Wetlands impacts could be hundreds of miles from the mitigation site. Such distant mitigation cannot truly compensate for lost functions and ecosystem services, which "are not abstract or portable" (ELI, 2002). No bank should have such a large service area, and an entrepreneurial banker is left to ponder the preferential treatment for a public-sector competitor.

Woodbury Creek (New Jersey)

This is the poster child for a bad mitigation bank (Gardner and Radwan, 2005). The saga also illustrates the great authority that a federal bankruptcy court has. In 1995 New Jersey approved the bank, which was expected to enhance about 128 acres of wetlands, create another 38 acres of freshwater wetlands, and provide more than 18 acres of upland buffer. If all went well, the bank would generate almost 100 credits. After the early release and sale of about a third of the credits, however, all went badly. The site did not have soils conducive to wetland creation. Compounding the problem was that LandBank, the bank sponsor, inadvertently drained approximately 19 acres of wetlands when trying to create wetlands. (LandBank's consultant notified New Jersey Department of Environmental Protection [NJDEP] that some of the "existing wetlands on the site no longer appear to exhibit wetland hydrology.") To pay for the remediation work, NJDEP looked to LandBank's performance bonds, which are financial assurances (like insurance) that can be drawn upon when certain contingencies occur. LandBank, however, had stopped paying the premiums on the bonds, which had lapsed. Thus, there was no ready pool of money that could be drawn upon.

NJDEP then brought an administrative enforcement action against LandBank, and LandBank was ordered to restore 57 acres (a 3:1 ratio for the draining) and to pay $9,000 in penalties. The only problem, however, was that LandBank's controlling corporation was in bankruptcy. The federal bankruptcy judge viewed NJDEP as being in the same position as an unsecured creditor, which is not a prime position in a bankruptcy proceeding. Unsecured creditors receive typically little or no payment. The state order to restore and to pay a penalty was ultimately wiped away.

They're only in it for the money (and other criticisms of mitigation banking).

Mitigation banking has its share of critics (Sibbing, 2005). Some question whether we can truly restore wetlands (it is an evolving art) and whether the availability of wetland credits weakens the sequence of "avoid-minimize-compensate." The Corps may relax—or feel pressure to relax—the need to avoid wetland impacts if compensation is in place.

Some opposition to mitigation banking flows from discomfort with entrepreneurial banks. The motive of the entrepreneurial banker is not entirely pure. Bankers undertake this effort to make a profit. Some environmentalists (and regulators) find it difficult to support an enterprise with such a supposedly base purpose (Shabman et al., 1994). But it is unrealistic to expect people to invest in wetland restoration out of the goodness of their hearts. (The angels among us who would do so probably do not have the funds.) Thus, this alleged negative is actually another strength of mitigation banking. Because there is the possibility of a return on one's investment, the private sector is voluntarily investing in and working on wetland restoration, enhancement, creation, and preservation. Mitigation banking enlists unlikely supporters of wetland regulation, which is necessary to create the market for credits. For example, a former staffer for Senator Lauch Faircloth, a conservative Republican from North Carolina, now runs one of the biggest wetland and stream restoration businesses in the state. (One year the back of the company's T-shirts proclaimed, perhaps ironically, "Tree Hugger.") The growth of mitigation banks will lead to increased data and knowledge that will contribute to more effective compensatory mitigation projects in the future, whether they are part of a bank or not.

But there is a danger that agencies may focus too much on the financial well-being of mitigation banks and not enough on the bank sites themselves. Although mitigation banking was billed as advance mitigation, since the 1993 interim guidance the agencies have recognized that allowing the early release of some portion of the overall credits could be necessary to ensure the financial stability of a bank. The draft 1995 guidance stated that up to 15 percent of credits could be released early, but the final version dropped a specific percentage, noting that some commenters viewed the 15 percent figure as a floor and others saw it as a ceiling. Thus, the amount of early releases would be left up to the Corps to determine on an individual basis.

An Environmental Law Institute report (2002) suggested that the proposed (and abandoned) 15 percent figure became a de facto floor and that the tail might be wagging the dog. More than 90 percent of mitigation banks received early credit releases. Of the banks studied, the average percentage of the amount of credits released early was 42 percent—that is, more than two-fifths of credits could be sold prior to the banks meeting any ecological performance standard. While these early releases were typically contingent on the banker securing rights to and placing a conservation easement on the mitigation site, mitigation banking did not seem to provide functioning mitigation in advance of impacts.

In addition to these criticisms, mitigation bankers faced a challenge on another front: competition from in-lieu fee mitigation offered by government agencies and environmental organizations.

Chapter 8

In-lieu Fee Mitigation: Money for Nothing?

[W]hile the law [of competition] may be sometimes hard for the individual, it is best for the race, because it insures the survival of the fittest in every department.

—*Andrew Carnegie,* The Gospel of Wealth
and Other Timely Essays *(1889)*

In theory, mitigation bankers could hardly complain about competition. Mitigation banking, after all, was pitched as a market-based approach, and the market assumes competition. And they believed they delivered a quality product. If mitigation bankers were not providing advance mitigation, they were at a minimum providing more timely mitigation and were subjected to the MBRT interagency review process. What really concerned the mitigation bankers was not competition per se, but what they perceived as unfair competition. They believed that other mitigation providers, such as those operating in-lieu fee programs, were getting a relatively free pass from regulating agencies. In their view, in-lieu fee programs were not being held to the same standards and were undercutting the market for wetland credits (Urban and Ryan, 1999).

What is in-lieu fee mitigation?

Like mitigation banking, in-lieu fee mitigation is a form of third-party mitigation. And like mitigation banking, in an in-lieu fee scenario, the permittee writes a check to a third party and is then relieved from its compensatory mitigation obligations; the responsibility for the mitigation shifts to the in-lieu fee administrator. (It is called "in-lieu fee mitigation" because the permittee contributes money in lieu of doing the mitigation project itself.) The in-lieu fee funds are typically deposited in an account that a not-for-profit organization or government agency (or both) manages for environmental purposes. A significant difference between mitigation banking and in-lieu fee mitigation is a matter of timing (NRC, 2001). Whereas the mitigation bank site must be identified and protected (through a conservation easement or deed restriction) before mitigation bank credits can be sold, in-lieu fee funds are generally accumulated for future projects. Thus, another significant difference is that unlike mitigation bankers who must make a capital investment in the compensatory mitigation project (or find partners willing to do so), the in-lieu fee administrator bears no financial risk. The in-lieu fee administrator has no skin in the game.

Nevertheless, in-lieu fee mitigation does offer some potential environmental benefits (Gardner, 2000). In many cases, in-lieu fee mitigation is an improvement over traditional, permittee-responsible mitigation. Because the latter often fails, the regulatory agencies and the public ultimately receive little or nothing in terms of compensatory mitigation. At least with in-lieu fee mitigation there is a greater likelihood of success, especially if the money is used primarily for preservation activities. In-lieu fee mitigation also makes it easier for agencies to require compensatory mitigation for small impacts, where the permittee might otherwise not be required to provide anything. This can help to offset the cumulative impacts of minor development activities.

Perhaps the most effective in-lieu program is the Virginia Aquatic Resources Restoration Trust Fund run by The Nature Conservancy in Virginia. In accordance with an agreement with the Corps' Norfolk District, the Conservancy accepts and pools money from small development projects to conduct larger compensatory mitigation projects on a watershed basis. From 1995 to 2009, the trust fund accumulated $53.4 million from permittees (and some violators who paid into the fund to settle claims), which generated $4.5 million in interest. The Corps approved the expenditure of $37.5 million for 108 mitigation projects, as well as $3.5 million for Conservancy staff, equipment, and overhead expenses. The reported ecological results have been impressive, as detailed in the summary of impacts (table 8-1) and the summary of aquatic resources restored and protected (table 8-2).

Not only has the fund provided for the restoration of wetlands at a high ratio (2.5:1 acres for nontidal wetlands and almost 9:1 for tidal wet-

TABLE 8-1. Summary of impacts, mitigation payments, and funds authorized during 1995–2009.

Resource Type	Impacts	Mitigation Payments ($)	Authorized Funds ($)
Nontidal Wetland	240.85 acres	20,370,100	14,120,900
Tidal Wetland	2.612 acres	628,600	648,000
Stream (pre-USM)*	163,428 linear feet	24,970,400	21,988,500
Stream (USM)	18,299 linear feet	7,454,000	782,600
Totals		**53,423,100**	**37,540,000**

Source: Virginia Aquatic Resources Trust Fund Annual Report—2009.
*USM refers to the Unified Stream Methodology, an assessment technique developed by the Corps and the Virginia Department of Environmental Quality.

TABLE 8-2. Program-wide leverage during 1995–2009.

Resource Type	Impacts	Restored	Protected
Nontidal Wetland	240.85 acres	612 acres	3,968 acres
Tidal Wetland	2.612 acres	23.4 acres	543 acres
Stream	181,727 linear feet	52,294 linear feet	668,164 linear feet
Upland/Riparian Buffer	N/A	259 acres	5,345 acres
Additional Protected	N/A	N/A	10,027 acres
Totals	**243.46 acres**	**894 acres**	**20,777 acres**
	181,727 linear feet	**52,294 linear feet**	**720,458 linear feet**

Source: Virginia Aquatic Resources Trust Fund Annual Report—2009.

lands), but it has preserved (through acquisition, conservation easement, or deed restriction) more than 4,000 acres of wetlands. With impacts of a little more than 240 acres, that is a ratio of greater than 16:1. The fund's activity in this regard recalls an old refrain about The Nature Conservancy—they protect the environment the old-fashioned way: they buy it.

But the Virginia Aquatic Resources Restoration Trust Fund is not representative of all in-lieu fee programs. No other program appears to approach the magnitude of these ratios; indeed, as we will see, the Corps often has difficulty tracking the money and how it is used. Which, of course, leads us to the question: What is the legal authority of the Corps to approve cash contributions as compensatory mitigation in the first place?

The legal status of in-lieu mitigation (the early years)

As was the case with mitigation banking, in-lieu fee mitigation lived in the shadows of the law. Not expressly condoned (or prohibited) in the Clean Water Act or in the Corps' and the EPA's regulations, in-lieu fee mitigation was first discussed in guidance documents. An added complication, however, is the legal constraint on federal agencies accepting funds from outside the normal appropriations process.

In chapter 2, we noted that Congress has the power of the purse over executive branch agencies such as the Corps and the EPA. The Constitution assigns to the Congress the power to raise revenue, and "[n]o Money shall be drawn from the Treasury, but in Consequence of Appropriations made by Law." Congress therefore exerts control over agencies by controlling their funding levels and directing how the funds may be used. If, in addition to those funds already appropriated, an agency could raise funds without congressional approval, such an action would diminish congressional oversight of the executive branch.

Several statutory provisions codify this constitutional principle, the most significant of which for in-lieu fee mitigation is the Miscellaneous Receipts Act (MRA). The MRA states the general rule that "an official or agent of the Government receiving money for the Government from any source shall deposit the money in the Treasury as soon as practicable without deduction for any charge or claim." As the GAO has emphasized, the MRA clearly means "any money an agency receives for the government from a source outside of the agency must be deposited in the Treasury" (GAO, 2004). An agency cannot operate a slush fund, no matter how well intentioned the endeavor might be.

Accordingly, constitutional and statutory principles prevented the Corps from accepting in-lieu fee payments directly. If the Corps did receive a check, the funds would go the Treasury and could not be used for a mitigation project unless Congress approved the expenditure (through an appropriations act). To get around this hurdle, the Corps arranged for the funds to go to a not-for-profit organization or a state or local agency (if those agencies had the authority to accept funds under state law).

Initially, in-lieu fee mitigation was done on an ad hoc basis (Gardner, 2000). If on-site mitigation was not practicable, a Corps district might negotiate a payment to a conservation group to offset wetland impacts. One of the earliest formal in-lieu fee programs can be found in a general permit issued by the Corps' Vicksburg District, which covers parts of Mississippi, Louisiana, and Arkansas. General permits authorize (or are supposed to be limited to) minor activities that result in minimal environmental impacts, and a 1987 Vicksburg District general permit authorized hydrocarbon exploration activities that affected up to an acre of wetlands. For each such activity (such as creating a gas well in a wetland), the permittee was required to donate $200 to a conservation organization or agency. While the Vicksburg District could be lauded for requiring compensatory mitigation for activities authorized by general permit (a rare occurrence at the time), it also established a precedent in which in-lieu fee payments failed to cover the full costs of offsetting permitted impacts.

In-lieu fee mitigation made its first appearance in national guidance in one paragraph of the 1995 mitigation banking guidance. In-lieu fee mitigation was defined as an arrangement in which "funds are paid to a natural resource management entity for implementation of either specific or general wetland or other aquatic resource development projects." The agencies contrasted in-lieu fee mitigation with mitigation banking, noting that in-lieu fee mitigation does not "typically provide compensatory mitigation in advance of project impacts" and that "such arrangements do not typically provide a clear timetable for the initiation of mitigation efforts." Nevertheless, the 1995 guidance observed that the Corps could approve in-lieu fee mitigation if there were "adequate assurances of success and timely implementation." The guidance called on the Corps to enter into formal agreements, similar to mitigation banking instruments, with in-lieu fee administrators.

Individual Corps districts, such as the Chicago and Buffalo Districts, began to promulgate their own guidance on the use of in-lieu fee mitigation. (This is an important point for those trying to understand agency behavior or locate applicable rules; sometimes the critical guidance documents emanate from an agency's local offices.) With a nod from Washington and procedures spelled out in individual districts, the number of in-lieu fee mitigation programs began to grow. So did the criticism and concerns.

"Educational" mitigation

One concern was that the in-lieu fees were not being spent on projects that would result in on-the-ground mitigation (Gardner, 2000). Instead, sometimes the ready money found its way to pay for or supplement wetland education and research projects. For example, in 1998, the Corps' Portland District allowed approximately a half acre of fill for a sawmill on the condition that the permittee donate $10,000 to the Oregon Division of State Lands. The money was then used to establish a wetland park on county-owned, high-quality wetlands. The new park, which is adjacent to an elementary school, includes class staging, discussion, and viewing areas. While environmental education is a laudable governmental objective (and one that I obviously support!), you might question whether paying for a government project was appropriate compensatory mitigation. Moreover, this particular mitigation project did not seem to truly provide an offset, since the mitigation site was already on government-owned land and thus not likely to be subject to development pressures.

Another, more egregious instance came out of the Corps' Louisville District. In exchange for federal permission to fill five acres of wetlands for building construction, the permittee agreed to contribute $45,000 to the Louisville Zoo Wetlands Exhibit. Zoo visitors now can enjoy a wetland trail featuring a three-quarter-acre shrub swamp that is also used as an outdoor classroom. But subsidizing an educational exhibit through wetland destruction would seem to send a mixed message.

Conflict of interest: Agency as regulator and competitor?

If an agency is intertwined with an in-lieu fee fund, as is the case with some Corps districts (which approve expenditures) and state agencies (which may be the recipient of the monies), conflicts of interest may arise. The permit applicant knows full well that the Corps or a state regulatory agency has the authority to stop a proposed project outright by denying a permit. An agency can also cause a project to die by withholding or delaying a decision. While an agency may note that the in-lieu fee option is simply one of several choices available to the applicant, most applicants simply want to proceed through the permit process as rapidly and inexpensively as possible. An incentive thus exists for an applicant to please the regulators by making a contribution to the agency's favorite charity. And, at least in some instances, the agency's favorite charity arguably is itself. Take, for example, the Corps' Chicago District, which in 1997 entered into an in-lieu agreement with the Corporation for Open Lands, in which the Corps retained "full authority" to approve use of the in-lieu fee account. The Corps' partner was known and referred to as "CorLands," thereby conflating the permit-decision maker (Corps) with the recipient of the permit-tees' funds (CorLands).

A second possible conflict of interest involves the regulatory agency's relationship with mitigation bankers. If an agency operates or oversees an in-lieu fee fund, it may be seen as essentially competing with mitigation bankers for the same mitigation dollars. Of course, the agency also controls the demand for and supply of mitigation bank credits.

Conflicts of interest may also arise with members of the MBRT who comment on proposed mitigation options. For example, the Galveston District granted a permit for a housing development with 25.9 acres of wet-land impacts with the condition that the permittee contribute $300,000 to the National Fish and Wildlife Foundation (NFWF). The NFWF, which is a congressionally created, private, not-for-profit organization, would then use the money to acquire property in the Brazos River watershed. The per-mit anticipated that the NFWF would eventually transfer the land to the FWS to expand the Brazoria National Wildlife Refuge Complex. Interest-ingly, another mitigation option under consideration was to allow the per-mittee to purchase credits from the Katy-Hockley Mitigation Bank, but the Corps rejected that alternative based, in part, on the FWS's comments. A mitigation banker might find that it is competing not only against other mitigation providers, but against an agency's pet projects that have not re-ceived full funding through the ordinary appropriations process. The rela-tionships created by such in-lieu fee arrangements should at least raise the eyebrow of a government ethics advisor.

Timing in life is everything.

In-lieu fee mitigation is, by definition, typically after-the-fact mitigation. The compensatory mitigation project is done sometime after the permitted activities that destroy wetlands. One question about in-lieu fee mitigation has been how long after the fact would the mitigation project begin? When did in-lieu fee administrators have to start purchasing property or restoring degraded wetlands? At least initially, the guidance—or the ad hoc arrange-ments—offered a great deal of flexibility, which was good for the fund ad-ministrator, but not necessarily good for the environment.

A 2003 study by the Environmental Law Institute found that 58 of 87 in-lieu fee programs did "not require that the collected funds be spent in a specific time frame." For those programs that established some parameters, the time frames ranged from one to ten years, with three years as the average. Many of the agreements also allowed for extensions.

But they're the good guys!

The Corps and other agencies may have acquiesced in the great lag time between permitted impacts and compensatory mitigation projects funded by in-lieu fees because of the nature of in-lieu fees. It can take time for the funds from small projects to accumulate to provide sufficient capital to begin a restoration project, and permit applicants were not required to take the in-lieu fee option. But the nature of the organizations operating and administering the in-lieu fee programs also contributed to the relaxing of standards. The fund administrators were largely environmental groups and land trusts; they were not in the mitigation business to make a profit. Because their motives were pure (what could possibly go wrong?), they were not subjected to the same level of scrutiny as mitigation banks, which led to problems. The agencies had forgotten Ronald Reagan's advice to "trust, but verify."

Sometimes the Corps did not require in-lieu fee program proposals to go through the interagency MBRT process. Instead, an in-lieu fee organization simply had to enter into an agreement or a memorandum of cooperation with the Corps. These organizations were permitted to accept funds from violators and to focus on preservation, rather than restoration. (Neither of which is necessarily bad, but neither was typically available to mitigation bankers.) A common flaw in many in-lieu fee agreements was the pricing structure, which might not cover all the costs of the contemplated mitigation. Indeed, the very structure of some in-lieu fee programs, in which potential sites were not even identified at the time of accepting funds, creates this uncertainty. But if the in-lieu fee organization underpriced the costs of the future mitigation and was unable to deliver, most in-lieu fee agreements did not hold the organization legally responsible. Perhaps it was assumed that environmental organizations, with their hearts in the right place, would make everything okay.

Sometimes, however, the pure of heart are lousy managers. And not-for-profit organizations can go bankrupt just as for-profit entities do. The 2005 bankruptcy of The Environmental Trust (TET), a not-for-profit California corporation, offers an instructive (and frightening) lesson (Teresa, 2006; Gardner, 2008). TET acquired mitigation properties and was responsible for the long-term management of more than 4,600 acres.[1] In doing so, TET did not collect enough money from the permittees to establish an endowment sufficient to cover its costs. Due to this underfunding (and commingling its endowment with its operational fund), in 2005 TET was forced to file for Chapter 7 bankruptcy—which will result in TET's liquidation. The bankruptcy court approved a plan to offer TET's mitigation properties (along with the long-term stewardship responsibilities) to a number of interested parties, including state and federal agencies and other environmental organizations. But these entities would essentially be accepting an unfunded mandate. If no one steps up to take responsibility for the long-term management of these sites, the conservation easements might be extinguished by the bankruptcy court. (Never underestimate the power of a bankruptcy court to change the rules.) The TET saga illustrates that agencies should not give environmental groups a free pass. These organizations, however well-intentioned, should be held to the same standards and monitored just like for-profit entities.

The 2000 in-lieu fee guidance

Responding to the concerns about in-lieu fee mitigation, the Corps, the EPA, FWS, and the National Oceanic and Atmospheric Administration (NOAA) issued a nine-page guidance document in November 2000. The guidance (which was not put out for public notice and comment in the *Federal Register*, but developed through a stakeholder workshop) sought to clarify the appropriate use of in-lieu fee mitigation. As in all compensatory mitigation documents, it emphasized the sequence of "avoid-minimize-compensate" and that in-lieu fee mitigation should not be used to avoid the avoidance requirement.

The guidance attempted to rein in in-lieu fee mitigation in a number of ways. If in-lieu fee mitigation was to be used to offset the larger impacts associated with individual permits, then the in-lieu fee agreement must be reviewed by the interagency MBRT, thus providing more oversight of the Corps and its partners. Even more significant was how the agencies dealt with in-lieu fee mitigation in the context of offsetting small impacts associated with general permits. For the first time (and following Congress's lead, albeit a bit late), the agencies expressed a preference for mitigation banks over in-lieu fee mitigation:

> [W]here on-site mitigation is not available, practicable, or determined to be less environmentally desirable, use of a mitigation bank is

preferable to in-lieu fee mitigation where permitted impacts are within the service area of a mitigation bank approved to sell mitigation credits, and those credits are available.

If the mitigation bank did not offer in-kind credits (e.g., provide credits for a freshwater marsh to offset similar impacts) or if the mitigation bank offered only credits derived from preservation, then the in-lieu fee option could be considered. But the guidance represented great progress for the mitigation bankers; it essentially called for the end of in-lieu fee mitigation on an ad hoc basis.

The 2000 guidance document placed additional constraints on the use of in-lieu fee mitigation. The funds could no longer be channeled to education and research projects. Preservation would be acceptable only in "exceptional circumstances." Furthermore, in-lieu fee organizations could no longer simply sit on the money. They needed to begin the restoration, enhancement, and/or creation projects in a timely manner. Specifically, the guidance stated that the in-lieu fee organization should acquire the mitigation site and begin initial physical and biological improvements (e.g., filling ditches, removing invasives) by the first full growing season after the funds were collected. If the Corps agreed to extend this time frame to the end of the second growing season, then the mitigation ratios needed to be adjusted upward to account for the delay.

The guidance was published in the *Federal Register*, but it contained the usual caveat:

> The guidance is not intended, nor can it be relied upon, to create any rights enforceable by any party in litigation with the United States. This guidance does not establish or affect legal rights or obligations, establish a binding norm on any party and it is not finally determinative of the issues addressed.

Nevertheless, if Corps districts implemented the guidance, in-lieu fee mitigation would be subjected to more rigorous oversight. But that would mean that Corps districts would need to keep track of and assess in-lieu fee programs, something that they had largely been unable to do.

Tracking in-lieu fee performance (or the lack thereof)

After the 2000 in-lieu fee guidance came out, two reports highlighted the woeful record-keeping of the Corps. A 2001 GAO investigation examined the use of in-lieu fee mitigation and found that the Corps simply did not exercise rudimentary oversight of many in-lieu fee programs. Eleven of seventeen Corps districts that authorized in-lieu fee mitigation programs claimed that the resulting mitigation more than offset the wetland acreage impacts caused by developers that contributed to the funds, but the GAO noted that the Corps' data did not support that assertion. Moreover, even though nine of seventeen Corps districts similarly claimed that their in-lieu fee mitigation programs had achieved no net loss with respect to ecological functions, half of those districts admitted that they had not even attempted to assess the quality of the mitigation. A 2003 study by the Environmental Law Institute yielded the same results. It found that tracking data were unavailable or incomplete for 45 percent of in-lieu fee programs.

Obviously, a change in culture (and perhaps a technology upgrade) was needed. But a guidance document is not the strongest vehicle with which to make that change. Indeed, even though the in-lieu fee guidance discouraged the use of in-lieu fees on an ad hoc basis, the Environmental Law Institute found in 2006 that "these one-time payments outside of approved in-lieu fee agreements continue to occur fairly frequently at the discretion of individual Corps districts." If the mitigation bankers and other opponents wanted to reel in in-lieu fee mitigation, something more would be required.

Yet, for wetland-related impacts, in-lieu fee mitigation was only one competitor of mitigation banks. No one forced a permit applicant to turn

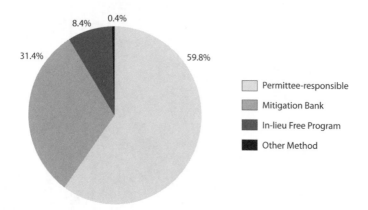

FIGURE 8-1. Wetland mitigation methods: proportion of required wetland mitigation nationwide (43,549 acres in fiscal year 2003) satisfied by permittee-responsible mitigation, purchase of credits from a mitigation bank, payment to an in-lieu fee program, and by other means. (Source: Wilkinson and Thompson, 2006).

to a mitigation bank or in-lieu fee program; they were only options that were offered to the permit applicant. Another alternative was for the permittee to perform the mitigation itself, which remained the most common method of satisfying compensatory mitigation requirements. The 2006 Environmental Law Institute study found that permittee-responsible mitigation provided almost 60 percent of wetland mitigation on an acreage basis (figure 8-1).

Permittee-responsible mitigation, however, was not held to the same standards as those applied to a mitigation bank. The permittee-responsible mitigation did not go through the formal interagency MBRT process, and permittee-responsible mitigation almost never contained provisions for long-term stewardship of the sites. And, of course, study after study demonstrated the shortcomings and failures associated with permittee-responsible mitigation. How could the playing field be leveled and standards harmonized for mitigation banking, in-lieu fee mitigation, and permittee-responsible mitigation? The answer was through a statute, regulations, and more guidance—although not necessarily in that order.

Chapter 9

Leveling the Mitigation Playing Field

> This final rule applies equivalent standards and criteria to all sources of compensatory mitigation, to the maximum extent practicable. It is not practicable to apply exactly the same standards and criteria to mitigation banks, in-lieu fee programs, and permittee-responsible mitigation, nor are the agencies required to do so.
> — *U.S. Army Corps of Engineers and U.S. Environmental Protection Agency, preamble to the April 2008 mitigation regulation*

"Leveling the playing field" had long been the mantra of entrepreneurial mitigation bankers. In their view, if all mitigation providers (banks, in-lieu fees, and permittees) were subject to the same standards, mitigation bankers would prevail (and profit) because they provided a superior product. Mitigation bankers called for equal treatment; what they got were "equivalent standards." As we will see, the Corps and the EPA have taken the position that equivalent standards do not necessarily mean equal or identical standards. But we will also see that the agencies embraced a preference for mitigation banking in the regulation, which seems to run counter to the concept of a level playing field.

An initial attempt at standards for permittee-responsible mitigation: The Halloween guidance

By the fall of 2001, the Corps and the EPA had jointly issued guidance for mitigation banks (1995) and in-lieu fee arrangements (2000), but had yet to provide detailed guidance for permittee-responsible mitigation, which accounted for the majority of mitigation acreage. The Corps, without consulting the EPA or any other agency, attempted to remedy this gap by unilaterally issuing a regulatory guidance letter (RGL) on Halloween in 2001 (Gardner, 2002). The EPA and environmental groups considered it more trick than treat. A blistering press release from NWF, Audubon, and others stated that the "arrogant" manner by which the Corps developed the new policy demonstrated a "complete lack of respect for the public" and "other federal agencies." On a substantive level, they deemed the new policy appalling: the press release dubbed it a retreat from wetlands protection, adopting an "anything goes approach" for mitigation that "signals the end of no net loss of wetlands within the regulatory program."

Were the criticisms of the Corps' Halloween guidance valid? Yes and no. The complaint about process was at least partially warranted, as the Corps issued the RGL without consulting other federal agencies. On the substance, however, the guidance itself attempted to adopt recommendations from the 2001 National Research Council report on compensatory mitigation. Despite the environmental groups' rhetoric, the guidance was hardly a clarion call for dump trucks to start filling wetlands. For the first time, Corps headquarters was trying to set standards for permittee-responsible mitigation. The guidance articulated what a permittee should include in a mitigation plan: baseline information of the site and mitigation goals, a work plan that explains how the site would be modified to achieve those goals, ecologically based success criteria to determine whether the goals had been met, monitoring and contingency provisions, financial assurances, and a description of how the site would be maintained and protected for the long term. But the Corps did so in a politically awkward (albeit legal) fashion, and process matters in Washington.

You could have at least called . . .

In response to criticism about the process, the Corps stated that it coordinates RGLs with other federal agencies when "the subject matter or the policies being provided require their involvement." The Halloween guidance was apparently not such a case. Attempting to diminish the importance of such guidance, the Corps insisted that "RGLs are used only to interpret or clarify existing regulatory program policy."

The Corps' explanation about the lack of notice to and communication with other federal agencies rang hollow on several levels. First, the Halloween guidance departed from the 1990 Corps-EPA Mitigation MOA (which had become more venerated over time) in several important respects. The Halloween guidance jettisoned the strong preference for on-site, in-kind mitigation in favor of a watershed approach in which on-site is not necessarily better than off-site mitigation and in-kind does not necessarily trump out-of-kind. The Halloween guidance also loosened restrictions on the use of preservation as mitigation, no longer limiting it to "exceptional circumstances." Putting aside for the moment whether these changes were good or bad, there is little doubt that they were significant deviations from the 1990 Mitigation MOA. Since the Corps and the EPA developed the 1990 Mitigation MOA together, the Corps should not have acted on its own. If it wished to amend the 1990 Mitigation MOA, the Corps should have discussed the matter with the EPA. If the Corps wanted to revoke the 1990 Mitigation MOA, it should have provided the EPA with six months written notice, as the document requires.

Second, there was much precedent regarding Corps-EPA coordination on RGLs, especially on guidance related to mitigation. A 1993 RGL that dealt with permit requirements for projects with minor impacts and with mitigation banking was actually two joint Corps-EPA memoranda to the field. A 1995 RGL that discussed individual permit flexibility for small landowners was also a joint Corps-EPA effort. Prior practice showed that "RGL" does not necessarily mean unilateral action.

The language of the Halloween guidance hinted at the frayed relationship between Corps headquarters and the EPA, as well as other resource agencies. With respect to agency roles and coordination, the RGL stated that the "Corps will often *choose* to coordinate proposed mitigation plans" with the EPA, the FWS, and other agencies (emphasis added). Not only is this statement wrong as a legal matter (the National Environmental Policy Act, the Endangered Species Act, and the Fish and Wildlife Coordination Act frequently compel coordination), but it seemed designed to offend (the Corps will speak with you when it deigns it appropriate), rather than to inform.

Lack of public input: Perhaps ill-advised, but legal

With respect to the lack of public input, the Corps correctly noted that it was not required as a matter of law to invite public comment before issuing RGLs and other guidance. The Administrative Procedure Act contains an exception from its public-notice-and-comment requirements for interpretive rules. Thus, the Corps and the EPA did not solicit public comment for the 1990 Mitigation MOA. Occasionally, however, the Corps and other agencies do seek public comment on mitigation guidance—as was the case with the 1995 mitigation banking guidance (through a notice-and-comment process) and the 2000 in-lieu fee guidance (through a stakeholders forum)—but they are not obligated to do so. As a policy matter, however, involving the public is a sound idea, especially when the guidance is likely to be controversial.

The Corps stated that it believed the Halloween guidance was noncontroversial, a laughable notion in light of the fact that almost everything in the section 404 program, especially mitigation, tends to ignite debate. Sometimes it is wise for agencies to go beyond the minimum of what the law requires.

Out of chaos comes order: The National Mitigation Action Plan

The Corps and the other agencies agreed to resolve the controversy over the Halloween guidance in the context of the National Wetland Mitigation Action Plan (MAP). Announced on Christmas Eve, December 24, 2002, the MAP was a joint effort by six agencies—the Corps, the EPA, NOAA, Department of Interior, Department of Agriculture, and Department of Transportation—to make better decisions about compensatory mitigation in light of the mounting criticism from the National Research Council, GAO, and others. The agencies agreed on seventeen action items to promote mitigation in a watershed context, improve accountability, clarify performance standards, and improve data collection and access. The first action item was to "clarify" the Corps' Halloween guidance, and the Corps issued a new RGL the same day that the MAP was launched. The Christmas Eve guidance went through interagency review and superseded the Halloween guidance, making several substantive changes in the process (such as only accepting preservation as mitigation in exceptional circumstances). With the controversy of the Halloween RGL out of the way, the agencies could turn to the other action items, many of which contemplated additional guidance related to difficult-to-replace aquatic resources, the role of preservation and vegetated buffers, and on-site/off-site and in-kind/out-of-kind mitigation.

Although as a legal matter the agencies could have issued the guidance without public involvement, they decided to use stakeholder forums to solicit input, as they did with the 2000 in-lieu fee mitigation guidance. The agencies, with the assistance of the Environmental Law Institute, held a series of workshops where interested representatives from government, environmental groups, academia, mitigation providers, and the regulated community came together to review progress on MAP action items and to provide comments on draft guidance. The Environmental Law Institute published reports on each of the forums and made the audio and PowerPoint presentations available to the general public on its Web site. In some respects, the innovative use of stakeholder groups to gather feedback on proposed guidance may be more useful than traditional notice-and-comment rulemaking. The stakeholder forums allowed parties to debate, argue, and discuss mitigation issues with one another, and agency representatives could ask follow-up questions (as could the other participants). Such exchanges can sometimes provide more insight into the real-life effect of policies than the formal letters submitted during a notice-and-comment rulemaking.

Congress (re-)enters the fray.

In 2003, while the agencies were working on guidance contemplated by the MAP, Congress directed them to take a different approach. Tucked into section 314 of the National Defense Authorization Act for Fiscal Year 2004 was the following provision:

(b) MITIGATION AND MITIGATION BANKING REGULATIONS—

(1) To ensure opportunities for Federal agency participation in mitigation banking, the Secretary of the Army, acting through the Chief of Engineers, shall issue regulations establishing performance standards and criteria for the use, consistent with section 404 of the Federal Water Pollution Control Act (33 U.S.C. 1344), of on-site, off-site, and in-lieu fee mitigation and mitigation banking as compensation for lost wetlands functions in permits issued by the Secretary of the Army under such section. To the maximum extent practicable, the regulatory standards and criteria shall maximize available credits and opportunities for mitigation, provide flexibility for regional variations in wetland conditions, functions and values, and apply equivalent standards and criteria to each type of compensatory mitigation.

Congress essentially told the Corps to level the playing field among mitigation providers and to apply equivalent standards to them—in a binding regulation, not mere guidance documents.

Congress ordered the Corps to issue the final regulations within two years—that is, by November 2005. That timeframe proved to be overly optimistic.

Proposed compensatory mitigation rule

Although section 314 was directed only at the Corps, both the EPA and the Corps agreed to conduct a joint rulemaking, which made sense given the structure of the Clean Water Act and the tradition (except for the Halloween guidance) of the agencies collaborating on mitigation policies. The proposed rule was published in the *Federal Register* in March 2006, four months after the deadline for the *final* rule. The delay could be attributed to resource constraints as well as the complexity of the issues. While much of the proposed rule incorporated or expanded upon existing guidance,[1] it also departed from existing policies in several important respects.

First, the agencies proposed to modify the definition of "mitigation bank." The proposal described a mitigation bank as

> a site, or suite of sites, where aquatic resources such as wetlands or streams are restored, established, enhanced, and/or preserved for the purpose of providing compensatory mitigation for authorized impacts to similar resources. Third-party mitigation banks generally sell compensatory mitigation credits to permittees whose obligation to provide mitigation is then transferred to the mitigation bank sponsor. The operation and use of a mitigation bank are governed by a mitigation banking instrument.

The careful reader will note what is missing: the definition no longer refers to mitigation banking as providing advance mitigation, mitigation prior to impacts. Instead, the agencies recognized (as the National Research Council and Environmental Law Institute pointed out) that mitigation banking did not typically provide fully functioning wetlands in advance of impacts. Accordingly, the agencies shifted the definition to reflect reality— that mitigation banks provided *performance-based* mitigation credits. Where permittee-responsible mitigation was a promise to perform the work in the future (the equivalent of a 100 percent early release of credits), mitigation banks only produced credits when they met certain performance-based milestones.

Second, the proposal suggested phasing out the use of in-lieu fee mitigation. In-lieu fee programs would have five years to modify their operations to comply with the requirements imposed on mitigation banks. After that grace period, in-lieu fee mitigation would no longer be a compensatory mitigation option. The agencies specifically requested comments, however, on whether to retain in-lieu fee mitigation, but under "specific, but somewhat different, requirements from mitigation banks." The agencies also solicited input about whether, if in-lieu fee mitigation was retained, a preference should be established for mitigation banks.

The Corps and the EPA provided the public a sixty-day comment period, which was extended another thirty days. The proposed rule generated approximately 850 distinct comments from the entire range of stakeholders—the regulated community, mitigation providers, state and local government—as well as the general public. Some of the submissions, such as the comments from the NWF and others, ran more than 100 pages. At the close of the comment period in June 2006, the agencies began the process of reviewing the comments. Meanwhile, the tensions between the proponents of mitigation banking and in-lieu fee mitigation spilled into the courtroom.

O'Hare Airport and the return of CorLands

The 2000 in-lieu fee guidance was meant to rein in the use of in-lieu fees, especially for individual permits. In-lieu fee mitigation was to be available mostly to offset small impacts authorized by general permits, and even then mitigation from mitigation banks was to be favored. The Corps' and the EPA's proposed rule advocated killing in-lieu fee mitigation. But like melaleuca or water hyacinth, in-lieu fees are a hardy species, difficult to keep in check and almost impossible to eliminate.

The O'Hare Modernization Project, approved by the Corps in 2005, illustrates this point. The airport expansion will fill 153 acres of wetlands and other waters, but only 97 acres of wetlands are jurisdictional under the Clean Water Act. The remainder is considered isolated and thus not waters of the United States under the reasoning of the *SWANCC* decision. As a practical matter (and consistent with interagency guidance with the Federal Aviation Administration), the compensatory mitigation for this project needed to be off-site. Although many airports are constructed on or near wetlands, birds and airplane engines do not mix (a point reinforced by the double-engine bird strike that caused Captain Sullenberger to land a US Airways jet in the Hudson River in 2009). It makes no sense to restore wetlands on-site if they are likely to attract wildlife hazards.

The permit applicant, the City of Chicago, proposed that part of the project's impacts be offset by 62 acres of mitigation credits from a mitigation bank. At some point, the Corps informed Chicago that it needed more compensatory mitigation (another 280 acres). The Corps suggested in a letter to Chicago that

> you seriously consider engaging an organization that has expertise in real estate transactions, planning, and executing wetland restoration programs to assist you in developing a potentially acceptable set of projects, if you are still interested in trying to meet your proposed schedule. One such organization well known for their work in this area in the Chicago region is Corporation for Open Lands (CorLands), an affiliate of the Open Lands Project. . . . We have worked with Cor-Lands on a number of programs over the past ten years, both having them as an administrator for a restoration program as well as a developer of a large wetland restoration program. . . . We have found them quite effective[.]

The Corps also stated that it was up to Chicago to decide how it wanted to proceed, but the message was clear: go with the Corps' trusted partner or else there will be delays. Chicago went down the path of least resistance and agreed to pay $26.4 million to CorLands to obtain another 280 acres of compensatory mitigation.

When the 2000 in-lieu fee guidance was issued, the Chicago District closed down CorLands for purposes of accepting in-lieu fee funds. It was reopened for the sole purpose of compensating for the O'Hare project. The National Mitigation Banking Association and several Chicago-area bankers sued, arguing that the $26 million in-lieu fee was inconsistent with the Clean Water Act and the 2000 in-lieu fee guidance.

Establishing standing was the first hurdle for the mitigation bankers in the O'Hare litigation, which became known as *National Mitigation Banking Association v. U.S. Army Corps of Engineers*. The plaintiffs had to show that they had suffered a particularized injury, that this injury was fairly traceable to the Corps' decision to approve the $26.4 million in-lieu fee program, and that a favorable court decision would likely redress the injury. The mitigation bankers relied on injury to their economic and competitive interests; they had made investments in wetland restoration and had credits to sell. Conceding the injury, the Corps suggested that they had no standing based on a lack of causation and redressability. The Corps argued that it was Chicago's choice not to purchase more credits from area mitigation banks "for reasons of its own." And even if the permit was remanded back to the Corps, who was to say whether Chicago would then choose to pur-

chase additional mitigation bank credits? The court rejected these assertions, calling them disingenuous. The record clearly showed that the Corps
had steered Chicago toward CorLands, and if the permit were reversed, it
was likely that Chicago would purchase at least some additional credits
from mitigation banks. But while the mitigation bankers prevailed on the
issue of standing, they did not fare so well on the merits.

In its 2007 opinion, the court observed that the case was "essentially an
attack on the use of in-lieu fees for any individual permit." Although the
court noted that the Corps agreed "that mitigation banks are preferable
over in-lieu fees" and that "the key to successful mitigation rests on appropriate site selection and approval of specific mitigation plans," it nevertheless upheld the Corps' decision. The court stated that in-lieu fee mitigation
was a legal option, and the Corps district had much discretion when impos-
ing mitigation conditions to individual permits. The court acknowledged
that the Corps' and the EPA's proposed regulation would phase out in-lieu
fee mitigation, but emphasized that this was only a proposal. It had yet to
be adopted, and thus the rulemaking process was the more appropriate forum to challenge the use of in-lieu fee mitigation. The court's opinion
drove home the point to mitigation bankers that the proposed federal regulations on compensatory mitigation must be finalized.

Reconsidering in-lieu fee mitigation

By late 2007, with the statutory deadline for the mitigation regulation long
since passed, the gestation period for the final rule remained uncertain. The
speculation was that the status of in-lieu fees was one of the issues causing
the delay. Despite the objections of mitigation bankers, in-lieu fee mitigation had politically powerful proponents. Homebuilders and other developers were supporters because it could be less expensive than mitigation
banks and responsibility for mitigation success still shifted to the in-lieu fee
entity. States and local governments that administered in-lieu fee programs
also supported the continuation of in-lieu fee mitigation. The Bush administration did not want to anger developers or infringe on state wetland programs and was therefore reluctant to eliminate the option.

Discussions among OMB, EPA, and Corps officials prompted the agencies to take a fresh look at in-lieu fee mitigation. Mitigation banks and in-lieu
fees typically had been viewed as subsets of third-party mitigation, and the
debate had focused on applying the same standards to each. But an in-lieu
fee entity is more akin to an environmental consultant hired by a developer

to perform an off-site mitigation project. As such, in-lieu fee mitigation could be considered more as a subset of permittee-responsible mitigation.

Recall that although a mitigation banker's currency is credits, what they are really selling is a release from liability. When a permittee purchases mitigation bank credits (with agency approval), it has satisfied its mitigation requirements. The responsibility for the success of the mitigation site and its long-term maintenance is transferred to the mitigation banker. That is the essence of third-party mitigation: a third party (the mitigation banker) becomes legally responsible for the mitigation site.

With permittee-responsible mitigation, the permittee typically hires an environmental consultant or engineering firm to perform the mitigation on-site or off-site. If the consultant fails to perform, it is the permittee that must answer to the regulatory agency. It is the permittee that remains responsible, legally and financially, for mitigation success.

The transfer of liability to a mitigation banker is appropriate, in part, because prior to the release and sale of credits, the banker must invest money up front, go through an interagency review process, acquire a site, place a conservation easement or other protections on the site, and satisfy additional performance standards. With permittee-responsible mitigation, typically the permittee promises to perform (or hire someone to perform) the mitigation in the future.

In-lieu fee mitigation is more like permittee-responsible mitigation: it is a promise to do good in the future. One might say that in-lieu fee mitigation is less performance-based and more promise-based. It lacks the essential element that gave regulators some confidence in banking: a site. Accordingly, if in-lieu fee mitigation was to survive, perhaps an in-lieu fee entity should be treated like an environmental consultant. In other words, when a permittee chose the in-lieu fee option, the permittee would remain responsible for the mitigation. If the in-lieu fee entity does not deliver as promised in a specified timeframe, the permittee will be on the hook to pay (a second time) for the mitigation.

Such an approach would alleviate at least two concerns about in-lieu fees: the lack of accountability and the underpricing of the true cost of mitigation. First, a party (the permittee) would be clearly identified as legally responsible for the success of the mitigation. Second, the ability to go after the permittee for additional funds should ensure that true costs of wetland mitigation are fully recovered. This, of course, would make in-lieu fee mitigation a less attractive option for permittees.[2]

Ultimately, the agencies decided to keep the in-lieu fee option and continue to treat it as third-party mitigation, in which the responsibility for the mitigation transfers to the in-lieu fee administrator. But the final rule also contained a reward for the mitigation bankers: a preference for mitigation bank credits enshrined in regulation.

Finally, the final rule emerges.

The Corps and the EPA issued their final rule on their Web sites on March 31, 2008. It was formally published in the *Federal Register* ten days later. The rules governing compensatory mitigation, long contained only in guidance documents, were now in regulations that had the force of law.

The regulation was divided into eight parts (figure 9-1). The first seven parts apply to all three types of mitigation, and the final part applies to third-party mitigation (mitigation banks and in-lieu fee programs). Each type of mitigation was not held to the same standards, however. Although the agencies claimed they had established equivalent standards, they emphasized that "equivalent" does not mean "equal." Let's examine how the regulation resolved some of the major debates about compensatory mitigation.

Sequencing and avoidance

The regulation reaffirmed the sequence of avoid-minimize-compensate (actually using the term "sequencing" for the first time in regulation). Compensatory mitigation proposals, whether from permittees, mitigation banks, or in-lieu fee programs, are not intended to dilute the initial duty to

MITIGATION FOR LOSSES OF AQUATIC RESOURCES
33 C.F.R. Part 332

332.1 Purpose and general considerations.
332.2 Definitions.
332.3 General compensatory mitigation requirements.
332.4 Planning and documentation.
332.5 Ecological performance standards.
332.6 Monitoring.
332.7 Management.
332.8 Mitigation banks and in-lieu fee programs.

FIGURE 9-1. Contents of compensatory mitigation regulation.

avoid wetland impacts. But how compelling is the duty to avoid wetland impacts? The regulation touched on the concept of avoidance in the context of difficult-to-replace wetlands. The 2001 National Research Council report (as well as many other commentators) had strongly recommended that permittees avoid impacts to wetlands such as fens and bogs that are difficult or impossible to restore. In response, the regulation stated:

> For difficult to replace resources (e.g., bogs, fens, springs, streams, Atlantic white cedar swamps) if further avoidance and minimization is not practicable, the required compensation should be provided, if practicable, through in-kind rehabilitation, enhancement or preservation since there is a greater certainty that these methods of compensation will successfully offset permitted impacts.

On the plus side, the agencies expanded the examples of difficult-to-restore wetlands. On the negative side, the regulation does not expressly call for greater avoidance of impacts to these wetlands. The term "further avoidance" might suggest that a permittee make greater efforts to avoid impacts, but even this vague exhortation is undercut by the modifier "not practicable." The definition of practicable allows considerations of cost to enter the equation, and then avoidance often becomes too expensive (not practicable) from the permittee's perspective. While the Corps and the EPA may issue additional guidance on this subject, we know that such guidance will not be binding on the individual regulator in the field. The strength of the avoidance requirement will depend on the strength of these individuals.

Equivalency in mitigation plans

The regulation calls for all mitigation providers to develop mitigation plans that contain a dozen common elements. Described by National Mitigation Banking Association counsel Peggy Strand as a "twelve-step program for mitigation," this requirement harmonizes the standards applied to permittees, mitigation banks, and in-lieu fee programs. The mitigation plan must include the following:

1. *Objectives*: What type (and how much) of wetland or other aquatic resource will be provided? How does the project contribute to meeting watershed needs?
2. *Site selection*: What are the factors (including watershed needs) that go into choosing the mitigation site?

3. *Site protection instrument*: Who owns the site? What will be the legal mechanism (e.g., conservation easement) to protect the site over the long term?

4. *Baseline information*: What are the ecological characteristics of the mitigation and impact sites?

5. *Determination of credits*: How are credits (improvement of functions at the mitigation site) provided?

6. *Mitigation work plan*: How will the compensatory mitigation project be accomplished? What are the construction details?

7. *Maintenance plan*: How will the mitigation site be maintained after initial construction?

8. *Performance standards*: What are the ecologically based performance standards that the site must meet to demonstrate that it is achieving its objectives?

9. *Monitoring requirements*: How will the site be monitored to ensure it is meeting the performance standards?

10. *Long-term management plan*: After the performance standards have been met, how will the site be sustained for the long term? What financial mechanisms will be in place?

11. *Adaptive management plan*: What is the strategy to respond to adverse changes in the mitigation site's condition?

12. *Financial assurances*: What are the financial mechanisms (such as bonds, letters of credit) in place to ensure that the construction will be completed and meet the performance standards?

Although many of these items could previously have been found in guidance documents, the regulation codifies them into firm requirements. Furthermore, the regulation now requires that the public notice of permit applications must include details about how compensatory mitigations would be provided. Similarly, public notice must be given for proposed mitigation banks and in-lieu fee programs. Yet a big difference in the treatment of mitigation banks remains: their credits cannot be used until a mitigation site has been acquired. The administrative and performance standards to create mitigation credits may be at least roughly equivalent, but the "criteria for use" of credits are not.

Nonequivalency in the timing of the use of mitigation credits

Under the 2008 regulation, mitigation banks, in-lieu fee programs, and permittee-responsible mitigation all produce mitigation "credits"—which

the regulation defines as "a unit of measure (e.g., a functional or areal measure or other suitable metric) representing the accrual or attainment of
aquatic functions at a compensatory mitigation site." A credit is how we describe the delta, the difference between the initial conditions at the mitigation site and the conditions after environmental improvements have been
made. A credit allows a development project that will adversely affect wetlands to proceed. While mitigation bank credits can only be used after certain milestones have been met, the credit market for in-lieu fee programs
and permittee-responsible mitigation is much looser.

Under the regulation, a mitigation bank can sell its credits only after it
has a mitigation banking instrument (containing the twelve-step mitigation
plan) approved by an interagency review team (IRT). It then must secure
the physical site and obtain appropriate financial assurances. At that point,
the Corps will allow "an initial debiting of a percentage of the total credits
projected at mitigation bank maturity." Additional releases of credit are authorized when the mitigation bank meets specified performance standards,
but the regulation states that the "credit release schedule should reserve a
significant share of the total credits for release only after full achievement of
ecological performance standards." While the regulation does not define
what constitutes a "significant share," the preamble to the rule notes that it
"does not necessarily mean a majority," but rather "a proportion of projected credits that will provide the sponsor with a significant incentive to
complete [the] . . . project and ensure that all performance standards are
achieved." The release of mitigation bank credits is tied to improvements at
the mitigation site itself, beginning with its acquisition.

The regulation also mandates that an in-lieu fee program have an interagency approved instrument before credits can be used. The mitigation site
need not be secured, however. The Corps may authorize "a limited number
of advance credits available to permittees" once the instrument is approved.
The regulation provides that the in-lieu fee program must obtain the mitigation site and complete initial improvements "by the third full growing
season after the first advance credit" has been sold, although the Corps can
grant additional time. The amount of advance credits, which again is ultimately left to the discretion of the Corps, should be based on the in-lieu fee
program's compensation planning framework (a watershed approach to selecting mitigation sites), the sponsor's past performance related to wetland
projects, and the financing necessary to begin the compensatory mitigation
projects. Once mitigation projects are implemented on the ground and
meet performance standards (thereby paying back the initial release of advance credits), the in-lieu fee program may receive additional advance cred-

its to sell. Thus, in-lieu fee mitigation is structured to operate as after-the-fact mitigation, with a limited but revolving fund of advance credits.

If in-lieu fee mitigation allows a "limited" amount of advance credits, then permittee-responsible mitigation can be seen as allowing a 100 percent release of advance credits. Once the mitigation plan is approved by the Corps (not the IRT) and the permit issued, the permittee can proceed with the development project. In essence, all the credits from the proposed mitigation have been debited, but the mitigation project has not yet commenced.

You can think of the different types of mitigation providers as individuals seeking loans. Some homeowners receive a traditional, fixed-rate mortgage after putting down 20 percent, and other homeowners take out a subprime mortgage with no money down. The former (akin to a mitigation bank that has acquired a mitigation site) is the more secure investment; the latter (in-lieu fee and permittee-responsible mitigation) might very well meet its payment obligations (or pay back the advance credits), but it is more of a gamble. All in all, however, these new procedures detailed in a regulation are an improvement over the old approach with permittee-responsible mitigation, which often was the equivalent of no-doc loans.

The mitigation hierarchy

What would equal standards applied to all mitigation providers look like? If the Corps and EPA held in-lieu fee programs to mitigation banking standards (no advance credits without a site), most in-lieu fee programs would shrivel up and die unless the sponsors became willing to invest their own capital up front. And development would screech to a halt if a permittee were required to have mitigation in place before proceeding with their housing development or big box store. While good from the environmental perspective, it is an academic pipe dream.[3] At the other end of the spectrum, what if mitigation bankers simply followed the standards of permittee-responsible mitigation? Credits would be bought and sold with little assurance that the promised mitigation would come to fruition. So, the political realities dictated that the Corps keep all three mitigation options and acknowledge that they could not be treated equally.

But the Corps and EPA recognized the value of mitigation banking: larger, ecologically valuable parcels, more rigorous scientific and technical reviews, advance site acquisition, project-specific planning, and some level of financial security. If they could not ratchet up the standards to apply to all types of mitigation, they did the next best thing. The agencies established a preference for credits from mitigation banks.

Specifically, the regulation announced a new mitigation hierarchy. At the top are mitigation bank credits, followed by in-lieu fee mitigation credits, permittee-responsible mitigation under a watershed approach, on-site and in-kind permittee-responsible mitigation, and off-site and/or out-of-kind permittee-responsible mitigation (figure 9-2). The Corps must consider compensatory mitigation options in this order, giving mitigation bank credits preference. It is only a preference, which can be overridden if the Corps determines a nonbank option can provide better environmental returns.

Even though in-lieu fee mitigation was not vanquished, mitigation banking won the battle decisively. Mitigation banking is king, and it's good to be the king.

But is the compensatory mitigation regulation good for the environment?

The mitigation bankers will benefit from the new regulation, but will the environment? We return to the old standard law school answer: It depends (and it is also too soon to tell). Here it depends on implementation at the Corps district level. Will the districts require more avoidance of difficult-to-replace wetlands? Will the districts follow the mitigation banking preference? Will they incorporate watershed planning into permit decision mak-

---

PREFERENCE HIERARCHY FOR COMPENSATORY MITIGATION

33 C.F.R. 332.3(b)

1. Mitigation bank credits
2. In-lieu fee program credits
3. Permittee-responsible mitigation under a watershed approach
4. On-site and in-kind permittee-responsible mitigation
5. Off-site and/or out-of-kind permittee-responsible mitigation

---

FIGURE 9-2. Compensatory mitigation hierarchy.

ing? Will they require all types of mitigation providers to establish permanent trust funds to care for the mitigation sites in the long term? The responses will depend, in large measure, on what Corps headquarters chooses to emphasize. As noted earlier, traditionally Corps SOPs have encouraged regulators to devote their attention and energies to processing and issuing permits, not visiting and assessing mitigation projects. If that continues to be the message from headquarters, then that will continue to be the priority (or lack of priority) in the field. To change behavior at the district level, Corps headquarters must change how it evaluates Corps districts. If the Corps rewards regulators for upholding rigorous mitigation standards, we will see better mitigation results. If there is any lesson to be learned from the saga of mitigation banking, it is that people will follow the incentives.

Chapter 10

Wetland Enforcement: The Ultimate Discretionary Act

Let gentleness my strong enforcement be.
—*William Shakespeare*, As You Like It *(1623)*

Both the Corps of Engineers and the EPA have enforcement authority under the Clean Water Act. If you illegally fill in a wetland or violate the conditions of your section 404 permit, you could have to pay thousands of dollars in administrative penalties. Or the agencies could bring a civil action in federal court where you could be hit with even more in fines, potentially costing you millions. A court could order you to remove the offending structures (such as a house) and to restore the site. If your transgressions are bad enough, you might even face criminal sanctions, including prison time. The agencies have some heavy hammers that they can swing to protect wetlands and to force compliance with the law. But the decision about whether to pursue an enforcement action, and how heavy the hammer blow should be, is largely a matter of agency discretion. Despite the enforcement tools available, the agencies' approach to enforcement is relatively gentle—which is as the regulated community likes it.

Who is the lead enforcement agency?

Although Congress gave enforcement authority to both the Corps and the EPA, it did not specify how they would share this power. To eliminate this

159

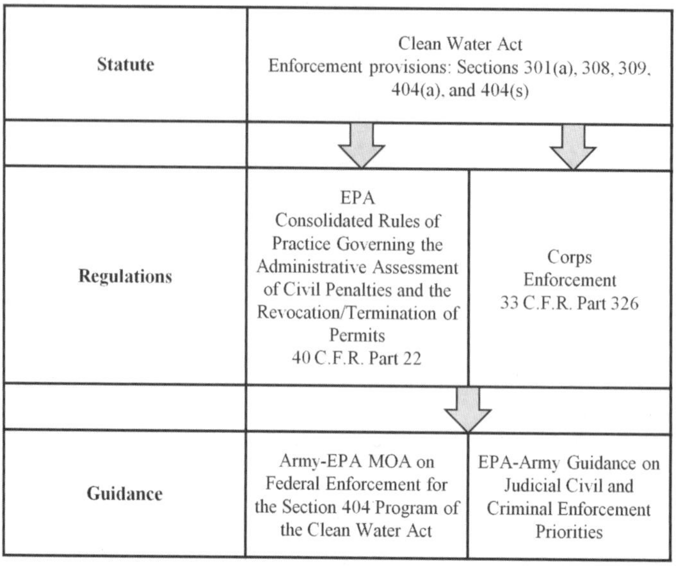

Statute	Clean Water Act Enforcement provisions: Sections 301(a), 308, 309, 404(a), and 404(s)	
	⬇	⬇
Regulations	EPA Consolidated Rules of Practice Governing the Administrative Assessment of Civil Penalties and the Revocation/Termination of Permits 40 C.F.R. Part 22	Corps Enforcement 33 C.F.R. Part 326
		⬇
Guidance	Army-EPA MOA on Federal Enforcement for the Section 404 Program of the Clean Water Act	EPA-Army Guidance on Judicial Civil and Criminal Enforcement Priorities

FIGURE 10-1. Legal and policy framework for enforcement of Clean Water Act section 404.

ambiguity and to make efficient use of the agencies' resources, the Corps and the EPA signed an enforcement memorandum of agreement (MOA) in 1989 that delineated their roles and responsibilities (see figures 10-1 and 10-2). The Corps agreed to be the lead agency for actions involving section 404 permit violations terms and conditions, which made sense since the Corps is the permit-issuing agency. For unpermitted discharges (i.e., when someone filled a wetland and failed to apply for a permit), the MOA left open which agency would be lead. It stated that the Corps would serve as lead agency, unless the case involved a repeat violator or a flagrant violation, and then the EPA would step in. The EPA would also be the lead agency, however, if it simply requested a particular case.

This division of responsibilities again seemed to make practical sense. The Corps has more wetland regulators out in the field than the EPA has, so let the Corps generally take the lead. But the MOA implicitly recognized that the EPA could intervene when it thought the Corps was not enforcing with sufficient vigor. The concern about the Corps' environmental commitment was why Congress gave the EPA such a significant role in the administration of the section 404 program in the first place. The Corps must apply EPA regulations when making permit decisions, and the EPA can veto a

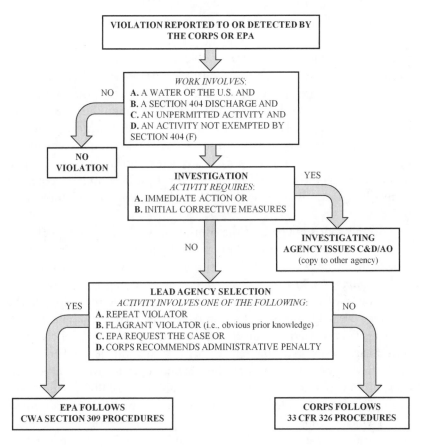

FIGURE 10-2. Division of Corps and EPA enforcement responsibilities. (Source: U.S. Army Corps of Engineers and EPA.)

Corps permit. In essence, the EPA also has a veto over the (non)exercise of the Corps' enforcement authority.

Every day is a new day: The continuing violation theory

Regardless of which agency takes the lead, the first step (if an investigation uncovers a continuing violation) is to tell the violator to stop what it is doing and remedy the situation. The Corps will do so with a "cease-and-desist" letter, while the EPA version is called an administrative order. The issuance of such orders is not determinative of liability; the alleged violator will still have an opportunity to contest the matter at some point. But the possible consequences of ignoring a cease-and-desist letter or administrative order are quite serious. If the agencies bring an action in federal court to enforce the order (judicial enforcement will be discussed shortly), the violator could face penalties of up to $27,500 per day per violation. Perhaps the amount might not immediately grab your attention, but the agencies have a "continuing violation" theory, in which they consider every day that unpermitted fill remains in place constitutes a new violation, which can add up. Do the math: under this approach, a violator who has illegal fill in place for one year is not looking at a maximum penalty of $27,500; the potential financial exposure is in excess of $10 million (365 days multiplied by $27,500).

Not all courts approve of the continuing violation theory, and even those courts that do accept it do not actually levy the maximum penalty. One of the first cases to adopt the continuing violation theory was *United States v. Cumberland Farms of Connecticut*. After receiving a cease-and-desist letter from the Corps to halt agricultural activities that were resulting in the discharge of dredged or fill material into the Great Cedar Swamp in Massachusetts, Cumberland continued landclearing operations. If the district court focused only on the discrete discharge events and the number of days in which they occurred (twenty), the maximum penalty would have been $200,000 (at the time the maximum penalty was $10,000 per day). Instead, the court concluded that "[a] day of violation constitutes not only a day in which Cumberland was actually using a bulldozer or backhoe in the wetland area, but also every day Cumberland allowed illegal fill material to remain therein." Ultimately, the court fined the company $540,000, although if Cumberland conducted the required restoration actions, it would be reduced by $350,000. Nevertheless, a precedent was established, and violators now had to face uncertain limits on their liability.

The continuing violation is also important in two other respects. First, it effectively eliminates a statute of limitations. The government (or citizen) must bring a case against a violator within five years of the violation; otherwise, the legal action is time barred. But if an unpermitted fill is a new violation every day it remains in place, then the five-year statute of limitations is never reached. The clock never starts running. For example, in *United States v. Reaves*, a Florida property owner illegally filled 17 acres of Alligator Creek and adjacent wetlands. The Corps did not discover the illegal fill for more than eight years, when it issued a cease-and-desist letter. Almost five years later—now thirteen years after the fill was put in place—the Corps went to court seeking civil penalties and a judicial order for the property owner to restore the site. The property owner admitted that the unpermitted fill had violated the Clean Water Act, but pointed out that the time to bring the lawsuit had long since passed. But the court adopted the continuing violation theory, and thus the government could proceed with its case.

The continuing violation approach also assists with citizen suits, as we will discuss in more detail shortly. Recall from our administrative law discussion in chapter 2 that in *Gwaltney v. Chesapeake Bay Foundation*, the Supreme Court held that a citizen may be able to bring a lawsuit against a Clean Water Act violator, but only for ongoing violations. If a defendant could argue that once the fill was in place, the violation was complete and no longer ongoing, then as a practical matter there would be no citizen suits for section 404 violations (except in those rare cases where a concerned group or person caught the discharger in the act). With the continuing violation theory, however, the private plaintiff could simply point to the discharge site and note that the illegal fill or dredged material was still in place, and thus was an ongoing violation.

Hobson's choice: No pre-enforcement review of administrative orders

If you are a recipient of a Corps cease-and-desist order or an EPA administrative order, your initial inclination might be to go on the offensive and sue the agencies. Perhaps you think that the area that you are discharging into is not a "water of the United States" for purposes of the Clean Water Act. Or perhaps you think that the activity that you are engaged in is exempted from Clean Water Act jurisdiction. You want your day in court to establish that what you are doing is perfectly legal. You may have the opportunity—but not right away.

Typically when you sue an agency, you can only challenge "final" agency action. A court does not want to get involved too early in agency deliberations. First, it would be disrespectful to a coequal branch of government. Second, it might be a waste of judicial resources. Once the agency makes a final decision, the prospective plaintiff might be satisfied and the need for a lawsuit evaporates. Only when the agency makes a final decision can the action be "ripe" for judicial review.

Consider what a cease-and-desist letter or an administrative order really is. On the one hand, it is a threat: stop what you are doing or you could be fined thousands (maybe millions) of dollars. On the other hand, it can also be viewed as a request: stop what you are doing and apply for a permit.

If you apply for a section 404 permit, the Corps might grant it. If the Corps grants it (or determines it has no jurisdiction), then the whole issue goes away. So courts are reluctant to intervene if the agency has merely issued an administrative order.

But the recipient of the order faces an uncomfortable choice: stop your activities and apply for the permit, or continue with your activities and take your chances that you will prevail when the Corps or the EPA takes you to court. Neither option is attractive. Suspending development operations can be very expensive, especially if equipment and workers are idled for months, and banks and investors are looking to be repaid. But defending against an enforcement action will also be quite costly even if you prevail. Your attorney fees cannot be shifted to the government. And if you lose, you are facing thousands of dollars in penalties and restoration costs.

Fortunately for violators (but not so much for the environment), the agencies prefer to settle. Their enforcement policies tend to emphasize bringing violators into compliance rather than punishing them to set an example. The willingness of the Corps to consider issuing section 404 permits to allow illegal discharges to remain in place illustrates this point.

After-the-fact permits: All is forgiven.

One of the Corps' enforcement options (and the most popular from a violator's viewpoint) is the after-the-fact permit. It is just what it sounds like: the applicant seeks a permit *after* the discharge has already occurred. Corps regulations state that an after-the-fact-application is to be processed in the same manner and subject to the same standards as an ordinary application. But of course an after-the-fact permit application complicates the Corps' review. The state of the wetlands prior to the discharge may not be well documented, and thus it may be difficult to assess the wetland functions already lost at the impact site. Moreover, how does a regulator conduct an alternatives analysis when the site has already been chosen? Corps regulations suggest that an after-the-fact permit should not be considered if permits or other legal authorization for the project has already been denied. If you are told no, you cannot go ahead with your project anyway and then seek forgiveness. Similarly, the Corps will not accept an after-the-fact permit application if the site is the subject of an ongoing enforcement action by other federal, state, or local agencies. Nor, the Corps regulations advise, are after-the-fact permits appropriate when the violation is willful, repeated, flagrant, or causes a substantial environmental impact. But it is up to the Corps to decide whether to accept an after-the-fact permit application, and traditionally the Corps has been most accommodating. Indeed, Professor William Rodgers refers to the Corps' approach as a "policy of mass amnesty."

The permit statistics, especially in the early years, suggest that there is much truth in the old saying that it is better to ask for forgiveness (after the fact) than to seek permission (before the discharge). A 1977 GAO report found that in four of five Corps districts surveyed unauthorized discharges were resolved through after-the-fact permits 87 to 99 percent of the time (Rodgers, 2009). A 1988 GAO investigation reviewed 87 sample violations and found that 36 cases were resolved through voluntary restoration of the sites and 25 (about 28 percent) were granted after-the-fact permits. The Corps "seldom required" compensatory mitigation for these after-the-fact permits, and the agency rarely invoked its authority to impose civil or criminal sanctions through the courts, even for serious violations. In the 1990s, the number of enforcement actions resolved through after-the-fact permits was estimated to be about 30 percent. It is difficult to say with precision how often after-the-fact permits are relied upon today. (In May 2010, the most recent statistics posted on the Corps Web site dated back to fiscal year 2003 and did not provide a category for after-the-fact permits.) What can be said with a high degree of confidence, however, is that the vast majority of unauthorized discharge cases are resolved either through the violator's voluntary restoration of the site or through after-the-fact permits. The Corps' and the EPA's enforcement swords—administrative, civil, and criminal penalties—largely remain sheathed.

Administrative penalties: Adjudication by the agencies

For the unlucky few whose violations are not resolved through voluntary restoration or after-the-fact permits, the agencies must decide whether to proceed with an administrative hearing within the agency or to go to federal court. If they opt for the former and keep it in house, they have two choices: Class I or Class II penalty proceedings. Class II proceedings are for more serious violations, and the maximum penalties that may be assessed are greater. For Class I penalties, the violator may have to pay up to $11,000 per violation with a cap of $27,500 for all violations. For Class II penalties, the maximum is $11,000 per day per violation, but with a cap of $137,500. Consequently, the procedural safeguards for the alleged violators are also greater in a Class II penalty proceeding.

A Class I or Class II penalty proceeding is an adjudication. (Recall from chapter 2 that an adjudication is the application of rules to a particular set of facts.) The alleged violator must be given notice. The hearing must be conducted by a neutral person, a hearing officer or administrative law judge. Although that person is an employee of the agency, he or she must make an impartial decision based on the record, based on the evidence introduced at the hearing.

The Corps and the EPA have slightly different administrative procedures for Class I penalties. If the Corps is seeking Class I penalties, a hearing officer will make a recommendation to the district engineer. The district engineer can accept, modify, or reject the recommendation, but the district engineer's final decision must again be based on the record. At this point, we finally have final agency action, and the violator can go to U.S. District Court to seek a review of the administrative penalties.

Under the EPA's procedures, Class I hearings are typically held by a regional judicial officer, an EPA attorney. After a hearing, the regional judicial officer will make proposed findings, conclusions of law, and a final order with recommended penalties. The recommended penalties will become final unless a party appeals to the EPA's Environmental Appeals Board, which would then issue a final order. The penalties can be reviewed by a U.S. District Court.

The EPA's procedures for Class II penalties follow a similar model, except the proceedings are run by an administrative law judge. Another difference is that judicial review of Class II penalties is at the Court of Appeals level. Although the Corps has had the statutory authority to assess Class II penalties since 1987, it has not yet promulgated regulations, which must be done through notice-and-comment rulemaking, to establish the governing procedures. Clearly, augmenting its ability to assess administrative penalties is not high on the list of priorities for the Corps.

Civil penalties: Potentially real money, rarely invoked

Occasionally, the Corps and the EPA will exercise their enforcement muscles and go after the more egregious violators in court to seek civil penalties (money) and injunctive relief (a court order to do something or refrain from doing something). The agencies cannot bring these civil judicial enforcement actions by themselves, however. They must work with the Department of Justice and convince the local U.S. Attorney's office to file a case at the U.S. District Court. Given the other priorities of the U.S. Attorneys, civil cases generally will be brought only for intentional, repeated, or flagrant violations or where there is great environmental damage.

The civil penalties can reach $27,500 per violation per day, which will add up under the continuing violation theory. Moreover, there is no maximum for civil penalties. In addition, the injunctive relief that is sought—typically an order that the defendant restore the destroyed or damaged site—can be an open-ended expense. But (as we saw in *Cumberland Farms*), a court will sometimes reduce or waive the civil penalties if the violator restores the site. The saga of *United States v. Cundiff* is a prime example. It also shows just how far a person has to provoke the federal government before it lumbers into bringing a civil action.

In this case, the defendants ignored Corps cease-and-desist letters and EPA administrative orders and continued to excavate, clear, and drain about 200 acres of wetlands along tributaries of the Green River in Kentucky. (The sidecasting of the dredged material constituted the discharge.) At one point, one defendant admitted that "though he knew he needed a permit, he thought the Corps would never grant him one so he planned on digging his ditches anyway." After eight years of warnings and negotiations, a civil action was finally filed. The U.S. District Court found for the government and ordered the defendants to pay $225,000 in civil penalties. The court also held, however, that all but $25,000 would be suspended as long as the defendants completed a restoration plan.[1]

So in the end the defendants might pay only $25,000 for ignoring government orders for eight years and illegally destroying 200 acres of wetlands. Of course, they have restoration expenses and attorney fees as well. But even if they paid the entire amount of the civil penalty, they would still only be assessed approximately $1,125 per acre. That is ridiculously cheap, especially when you consider how intentional and flagrant their actions were. Compare that penalty with the cost of purchasing mitigation credits at $20,000 per acre (which is at the low end of the scale). The mitigation credits for 200 acres would be $4 million.

Yet sometimes a court will impose what may be seen as a harsh penalty. A Texas dentist was ordered to pay $65,000 in civil penalties for filling in one-fifth of an acre of wetlands on Galveston Island. Although the size of the violation was small, he was a repeat and intentional violator. He had settled a previous case for $15,000 and admitted to "telling his contractor he frankly did not care about the wetland designation." While the agencies rarely use their hammer, it can still sting when they do.

Settlements, supplemental environmental programs, and other payments

Despite the likelihood that a court will not impose the maximum civil penalties, many defendants do not want to roll the dice. They would rather settle and dispose of the matter by entering into a consent decree. A consent decree is an agreement between the defendant and the agencies that is approved by the court. Department of Justice regulations require that the public be notified and be given an opportunity to comment on proposed consent decrees. (To find them, go on to the *Federal Register* Web site and search using the terms "consent decree," "Clean Water Act," and "dredged or fill material.") The defendants that enter into consent decrees seem to pay a surcharge for certainty. For example, in 2009 in *United States v. Savoy Senior Housing Corporation*, to settle a claim of illegally filling streams and wetlands in Liberty Village, Virginia, the defendants agreed to pay $300,000 in civil penalties, spend about $250,000 to conduct on-site restoration, and purchase approximately $825,000 of stream and wetland mitigation credits in the watershed.

One benefit of settling for defendants is closure. Litigation is stressful, and a consent decree can conclude the matter (as long as its terms are complied with). Furthermore, with the cessation of litigation, attorney fees subside. It may also be possible for the defendant to garner some favorable publicity if it agrees to voluntarily fund local environmental projects. The agencies welcome these "supplemental environmental projects" (SEPs) and take them into account in settlement negotiations. Indeed, because of the local benefits, the SEPs might be more welcomed than greater civil penalties.

When a defendant pays a civil penalty, that money goes to the general fund at the U.S. Treasury. The money does not go the local U.S. Attorney's office, the Corps district, or the EPA region. If the agencies did receive these funds, they would be augmenting their congressionally approved appropriations and violating the Miscellaneous Receipts Act (which Congress enacted to ensure that agencies remain within the fiscal limits set by Congress). While the imposition of high civil penalties can serve as a punishment to the violator and act as a deterrence to others, it does nothing for the local environment and does not compensate for lost ecosystem services. An SEP, however, is a vehicle to make sure that the defendants' money stays in the area.

For example, to settle an EPA enforcement action related to its bridge building and repair work, Lunda Construction Company agreed to pay

$10,900 in penalties and to remove the Deerskin River dam in Wisconsin as a SEP. The dam removal, which cost about $59,000, would restore a free-flowing river, thereby enhancing fish habitat. A more substantial settlement involved the Material Service Corporation, a mining company that had destroyed 37 acres of prairie wetlands near the Des Plaines River in Illinois. The company agreed to pay $500,000 in civil penalties and contribute $7 million to CorLands to restore and preserve similar habitat.

Technically, because the funds do not go to the EPA or the Corps, SEPs are not considered an improper augmentation of their budgets. While some might argue that this is form over substance, since the agencies can effectively control how the funds are spent, there are few churlish critics.[2] The violator is making the payment voluntarily to make the enforcement action go away. The local recipients and beneficiaries of the funds are not going to complain. And the agencies are happy because resolution of an enforcement action results in a local, environmentally beneficial program, rather than sending dollars off to the Treasury.

Sometimes, however, the violator's conduct is so bad that financial penalties are not sufficient and environmental programs cannot wipe the slate clean. In these cases, the U.S. government can choose to bring a criminal prosecution.

Criminal penalties: Muddy jackboots?

It takes a certain amount of stubbornness on the part of violators, coupled with the belief that the rules do not apply to them, to trigger a federal prosecution for a Clean Water Act section 404 violation. John Rapanos, a key figure in the Supreme Court case that brought us the "significant nexus" test for waters of the United States, illustrates this perfectly. He wanted to build a shopping center, but was informed by state regulators that the site was a wetland and that he would need to apply for a permit. He hired Dr. Goff as a consultant who told him the same thing; between 48 and 58 acres of one site was jurisdictional. Mr. Rapanos responded by instructing the consultant to destroy his "report and map, as well as all references to Mr. Rapanos in [his] files . . . Mr. Rapanos stated he would 'destroy' Dr. Goff if he did not comply, claiming that he would do away with the report and bulldoze the site himself, regardless of Dr. Goff's findings." Mr. Rapanos began landclearing activities, which led the state to come back with a search warrant. The violation resulted in EPA issuing an administrative order (which was ignored) and a federal prosecution.

But a criminal prosecution for a wetland violation, even if the violation is intentional, is very controversial. It is not the type of environmental crime that ordinarily results in imminent danger to the public, as is the case with the improper disposal of some hazardous substances. The "pollutant" could be clean soil or sand, and the activity is taking place on private property. The defendants are businesspeople, contributing to the local economy, not common street thugs. Thus, even when a particular person deserves to be prosecuted, the agencies and the U.S. Attorney's office will tread very gingerly because they will be pummeled in the press and perhaps by Congress. That was certainly the case with John Pozsgai.

John Pozsgai was, by virtue of his background, a most sympathetic defendant. He was born in Hungary and opposed the communists. He came to America in search of freedom (and private property rights). He was a successful small businessman in Morristown, Pennsylvania, who wanted to expand his mechanic shop by filling in a mosquito-breeding nuisance of a dumping ground. And for this supposed high crime, EPA jackboots started a federal investigation that resulted in Mr. Pozsgai receiving a sentence of twenty-seven months in federal prison, a $200,000 fine, and an order to restore the site. At least this is how some newspaper columnists and members of Congress portrayed the matter. The reality on the ground was a little different.

Mr. Pozsgai knew the site was a wetland when he purchased it. Indeed, he negotiated a reduced price because it was wet. When he started filling the site, he received cease-and-desist letters from the Corps, which he ignored. The Corps obtained a temporary restraining order from the U.S. District Court, an injunction that ordered Mr. Pozsgai to stop filling the site. He ignored the court order and was held in contempt. He ultimately deposited more than 400 truckloads of fill material covering 4 acres of wetlands. The water from the former wetland flooded his neighbors' properties. Indeed, one neighbor permitted federal investigators to use his property to video the comings and goings of the dump trucks. A jury found Mr. Pozsgai guilty of forty counts of Clean Water Act violations.

Putting aside the fact that his illegal fill caused flooding and harmed his neighbors' properties, the prosecution was entirely justified as a matter of law. Mr. Pozsgai refused to obey lawful orders from federal agencies and a federal judge. (It is never a wise idea to ignore the order of a federal judge, especially when he will be sentencing you.) Yet the prosecution garnered much bad press, and the Corps and the EPA were criticized at congressional committee hearings for beating up on the little guy.

Shortly after the Pozsgai controversy in the early 1990s, the Corps and the EPA proposed a "Wetlands Enforcement Initiative" (Gardner, 1996). The initiative sought to "emphasize the Federal government's commitment to Clean Water Act section 404 enforcement, to generally educate the regulated community and the public at large about the requirements of the section 404 program and the importance of wetlands, and to publicize Clean Water Act violations." Corps districts and EPA regions were to identify planned or pending enforcement actions against egregious violators that would then be publicized. The proposed initiative was not particularly well received in Congress. (Senator Symms of Idaho invoked a columnist's description of the initiative as "a brutal display of naked power.") The agencies never went forward with the enforcement initiative, and criminal prosecutions are reserved for rare cases.

The least sympathetic defendants

One such rare case that deserves mention is *United States v. Lucas*. The defendants filled hundreds of acres of wetlands to develop Big Hill Acres, a subdivision in Vanncleave, Mississippi. Despite warnings from the Corps and the EPA that the project required a section 404 permit—and warnings from the Mississippi Department of Health that septic systems in saturated soils would result in a public health threat—the developers moved forward and sold the homes to hundreds of low-income families. When the septic systems failed, raw sewage began "to seep up from the ground and flow across the development" (U.S. Department of Justice, 2005). In some cases, the wastes backed up into people's homes. In addition to Clean Water Act violations, the defendants were charged with conspiracy and mail fraud (for claiming that the homes were habitable). In 2005, a jury convicted the defendants on all counts. Robert Lucas received the longest sentence of nine years in prison. Two other individuals were sentenced to serve eighty-seven months. Two corporations were also fined a total of $5.3 million.

Big headlines about civil and criminal monetary penalties can be misleading. Often civil fines are not collected. The defendants may be "judgment proof" (lacking sufficient assets), and a corporation might go bankrupt and liquidate. And if a business can pay, it can just pass those costs on to its customers or consumers. But prison time is one cost that cannot be passed on.

Citizen suits: Backing up the government

Enforcement, as we have seen, is a discretionary act. What happens if the agencies decide to exercise their discretion *not* to enforce? If the federal (and state) government is not diligently prosecuting unauthorized discharges, the Clean Water Act authorizes "any citizen" to file a lawsuit against the alleged violator. A citizen, in this context, is defined as "a person or persons having an interest which is or may be adversely affected." Recall from chapter 2 that to bring a citizen suit, the person must also meet the Article III constitutional standing requirements of injury-in-fact (which should be satisfied if his or her interests are adversely affected), causation, and redressability.

Before commencing the suit, the adversely affected person must provide notice to the EPA, the state, and the alleged violator. The notice offers the EPA and/or the state the opportunity to take enforcement action, if they deem it appropriate. If either does, then the citizen suit cannot proceed. The alleged violator is notified to give it the chance to remedy the situation to avoid litigation. If the alleged violator takes corrective action, and the violation is considered to be wholly in the past, then a citizen suit is barred. A citizen suit may pertain only to ongoing violations.

This brings us back to the continuing violation theory. If every day that the illegal discharge remains in place is considered a new violation, it simplifies the task of establishing an ongoing violation. The violator would need to remove the fill (and possibly restore the site) for the violation to be considered wholly in the past.

A citizen suit can seek injunctive relief and civil penalties. The injunction will typically be a court order to the violator to stop discharging without a permit and to restore the site. While the violator may also have to pay civil penalties, a prevailing plaintiff in a citizen suit will not receive any damages. If the court imposes civil penalties, that money is paid to the Treasury. A court does have the discretion, however, to order the violator to pay the prevailing plaintiff's litigation costs, including attorney fees.

Enforcement of permit conditions: A gap in citizen suits

Although a citizen suit can be brought for unauthorized discharges, one cannot be brought for violations of permit conditions, such as the failure to provide promised compensatory mitigation. Why can an action be brought for one violation but not the other? The answer can be found in the words of the statute.

Congress authorized the use of citizen suits in section 505 of the Clean Water Act. That section expressly says that a citizen suit can be brought for alleged violations of section 402 permits issued by the EPA. Environmental plaintiffs regularly file suits to enforce water quality standards contained in section 402 permits.[3] But Congress neglected to include a similar provision for section 404 permit conditions. It is silent on that point, but courts presume that the omission was intentional. Thus, courts have rejected attempts by plaintiffs to sue to force section 404 permittees to comply with their permit conditions. Enforcement of these permit conditions is entirely within the discretion of the agencies.

Enforcement of third-party mitigation providers: Does responsible mean liable?

Yet the Corps and the EPA may also face difficulties in enforcing permit conditions when the responsibility for compensatory mitigation is transferred to a third party, such as a mitigation bank or in-lieu fee program. The linchpin to third-party mitigation is the transfer of liability. Once the permittee writes the check (with the Corps' approval), then the permittee has satisfied its legal obligations with respect to compensatory mitigation. The mitigation banker or in-lieu fee sponsor is now responsible for providing the environmental benefits on the ground. But what if they fail to do so? Is the mitigation banker or in-lieu fee sponsor liable under the Clean Water Act? Can they be subject to administrative, civil, or even criminal penalties?

Surprisingly, the answer is probably not. The mitigation banker and in-lieu fee sponsor are not engaged in unauthorized discharges. They are also not the section 404 permittees. Thus, there seems to be no statutory hook to bring an enforcement action against a failing mitigation bank or in-lieu fee program. Corps and EPA regulations refer to third parties assuming "responsibility" for the compensatory mitigation. But just as "equivalent" does not necessarily mean "equal," "responsibility" does not necessarily mean "liability," at least for purposes of the Clean Water Act.

With the enforcement mechanisms of the Clean Water Act likely unavailable, the agencies must become creative when dealing with a mitigation bank or in-lieu fee that is not meeting its responsibilities. Of course, the agencies can halt the release or sale of additional credits and threaten the individuals associated with the mitigation bank or in-lieu fee program that they will not be approved to operate third-party mitigation projects in the future. But that may not be much of a penalty, especially if a large number of credits have already been sold and there are no environmental gains to show as a result. A case involving a mitigation bank in Kentucky suggests that the agencies might be able to rely on contract-related claims.

In 1999, the Corps' Louisville District approved the establishment of the Wetland Bank of Kentucky through a memorandum of agreement (MOA), signed by the bank sponsors, the Corps, and other federal and state agencies that were members of the MBRT. The MOA contemplated that the bank would progress (and credits would be released) through five phases: removing cattle from the site; grading and constructing a berm; planting wetland vegetation; removing and keeping invasive species under control; and semiannual monitoring. The bank was also supposed to place a restrictive covenant on the bank site. The Corps released 11.35 credits, which the bank sold for approximately $110,000. The mitigation bank only removed the cattle (phase I) and did nothing more.

After years of discussions, in 2005 the local U.S. Attorney filed a civil complaint against the operators of the Wetland Bank of Kentucky, alleging breach of contract, negligent misrepresentation, breach of duty of good faith and fair dealing, and unjust enrichment. The defendants settled the case by agreeing to pay $70,000 to the Kentucky Department of Fish and Wildlife Stream and Wetland Mitigation Trust Fund (*United States v. Hawkins*, 2006).

Because the action was resolved through a consent decree, the court never definitively ruled on the government's contract-related claims, in particular whether the MOA was an enforceable contract. The episode raises some interesting questions about the basic principles of third-party mitigation. At a minimum, the approach the government took suggests that third-party mitigation providers are not subject to Clean Water Act liability (even though what they are selling is essentially a release of liability for the permittees). It also shows that wetland-related enforcement is relatively gentle. The defendants received almost $110,000 in credit sales, and they only had to pay $70,000 to a trust fund seven years later. Taking attorney fees into account, perhaps the defendants broke even or only made a slight profit, which hardly seems like a just punishment. But at least they were held accountable and had to pay something, an outcome that does not frequently come to pass.

We now turn from mitigation bankers' and permittees' responsibilities to those of the federal government. We will examine the topic of regulatory takings, when the government is obligated to pay just compensation to wetland owners, and whether agencies are held accountable for actions that diminish the value of property.

Chapter 11

Regulatory Takings in the Wetland Context

> The general rule at least is that while property may be regulated to a certain extent, if regulation goes too far it will be recognized as a taking.
> — Pennsylvania Coal Company v. Mahon, *260 U.S. 393 (1922)*

Federal regulation of wetlands begins with the U.S. Constitution. To justify the assertion of authority over privately owned wetlands, Congress must rely on an affirmative grant of authority enumerated in the Constitution. As we have seen, the Interstate Commerce Clause provides Congress (and thus the Corps and the EPA) the principal gateway to wetland regulation. The Constitution, however, also supplies a check on this federal authority through the Fifth Amendment's prohibition on the governmental "taking" of private property without the payment of just compensation.

The Fifth Amendment does not prohibit the government from taking your property. The government can confiscate your lands and boot you off for a highway project, for a public park, and even for a private redevelopment project (*Kelo*, 2005). In the wetland context, the government might even take your property to establish a mitigation bank.[1] But when the government physically takes your property, it must pay you a fair amount of money for your loss.

Of course, the government can interfere with private property rights without physically confiscating or occupying someone's land. An agency such as the Corps or the EPA may merely regulate the activities that you can conduct on your property. You still own the title to the property, and you can still walk on the land (and you can still pay property taxes). You may not, however, be able to put the land to the specific uses for which you bought it. In such cases, the property owners may feel that the government has effectively taken their land just as if it had physically seized it. And the U.S. Supreme Court, in an opinion by Justice Oliver Wendell Holmes, long ago observed that sometimes a regulation may go "too far" and result in a taking of private property.

Although this concept of "regulatory takings" (as opposed to physical takings) is well recognized by courts, it is exceedingly difficult for a property owner to prevail in these cases. There are procedural hurdles that must be overcome just to get into court and have the case heard. Once in court, the tests that judges use to determine whether a regulatory taking has occurred are advantageous to the government. Frequently, the property can be put to some use, so the governmental restriction has not wiped out its entire economic value. Even the advent of mitigation banking (which could be an economically valuable alternative use) strengthens the government's hand. A property owner who has purchased the property after the enactment of the statute or regulation burdening the property will not be in an enviable position. If your intended use of the property would result in a nuisance (e.g., filling a wetland that causes flooding on the neighbor's property), the government will not have to pay you anything. Consequently, the government rarely loses regulatory takings cases, especially in the wetland context.

Nevertheless, the Corps is skittish about infringing on a permit applicant's private property rights. For example, in a 2009 case challenging the Corps' issuance of section 404 permits for limestone mining in South Florida, Judge William Hoeveler documented how the Corps worried about a potential takings claim and related litigation costs. The judge expressed concern that the Corps seemed to be considering irrelevant nonenvironmental factors (i.e., the risk of a takings claim), even though the mining operation might lead to the contamination of an aquifer that supplies drinking water to Miami–Dade County. Indeed, the threat of takings claims is one explanation for why the Corps denies so few section 404 permit applications. But is this fear about regulatory takings warranted?

Preliminary hurdles: Ripeness

Some property owners believe that if their property is classified as a wetland, then a taking has occurred. The U.S. Supreme Court, however, has expressly rejected this assertion. The Court has held on numerous occasions that the mere assertion of regulatory jurisdiction by a federal agency over private property does not amount to a taking. In *United States v. Riverside Bayview Homes* (discussed in chapter 3), the Court noted that the Clean Water Act's requirement that a property owner secure a section 404 permit for the discharge of dredged or fill material into a wetland is not, by itself, a taking. A takings claim is simply not yet ripe. Why not? Because, if the property owner applied for a permit, the Corps may very well grant it, as the permit statistics suggest.

Similarly, the issuance of a cease-and-desist order or administrative order does not, by itself, constitute a taking. What is a cease-and-desist order? To be sure, it commands the property owner to stop the discharges. But such administrative orders are not final agency action. Moreover, the violator might apply for and be granted an after-the-fact permit.

There are unusual cases in which an administrative order can result in a ripe takings claim. In *Laguna Gatuna v. United States*, the EPA claimed Clean Water Act jurisdiction over hydrologically isolated playa lakes in New Mexico and ordered a company to halt the discharge of oil field brine.[2] The company first tried to fight the administrative order, but as noted in chapter 10, there is no pre-enforcement judicial review of EPA and Corps orders. The company then claimed that the EPA's order amounted to a regulatory taking of its property. While the takings case was pending, the U.S. Supreme Court decided *Solid Waste Agency of Northern Cook County*, which invalidated the Migratory Bird Rule and undermined the EPA's authority over the playa lakes. The EPA then withdrew its administrative order, more than nine years after it was issued. In such an extreme case, the court found that the takings claim was ripe.

Still, the general rule is that a property owner must apply for, and be denied, a section 404 permit before a wetlands takings claim will ripen.[3] And a permit applicant cannot simply get frustrated with the application process and walk away; it must see the permit process toward its conclusion. Otherwise, if a permit application is incomplete or the applicant does not provide information the Corps or the EPA has requested, then the Corps might deny the permit application "without prejudice," which means two things. First, it means that the property owner can apply again for a permit in the future. Second, because the property owner can apply again, the Corps does not view a denial without prejudice as final agency action. Thus, a takings claim would not be ripe as a result of a denial without prejudice.

Consider the story of Pax Christi Memorial Gardens, a cemetery in Lake County, Indiana. For more than two decades the cemetery had been unsuccessfully trying to expand its site into an area known as Evergreen Memorial Park. In 1980, the Corps informed the cemetery that, because of the presence of wetlands, it would need a section 404 permit to fill Evergreen Memorial Park.[4] In 1983, the Corps denied the permit, reasoning that burying bodies was not a water-dependent activity; thus, the section 404(b)(1) guidelines presumed that alternative upland sites with less environmental impact were available. After a hiatus, the cemetery continued discussions with the Corps and applied again in 1995 and 1997. The Corps responded by requesting more information concerning the need for additional burial plots, the availability of off-site alternatives, on-site alternatives, and compensatory mitigation. When the Corps did not receive the information, it administratively withdrew the application. Pax Christi eventually brought a takings action, but the court held that the claim was not yet ripe. Despite the cemetery's twenty-year quest, the court noted that the Corps had not made a final determination: "when the government has *withdrawn* a § 404 permit application for lack of necessary information, there has not been a final decision for ripeness purposes." It can be tough to get into court to raise a regulatory takings claim.

Choosing a forum: U.S. District Court or the U.S. Court of Federal Claims

If you do apply for a section 404 permit and are one of the unlucky few who are denied, you have two primary judicial options.[5] If you want to overturn the Corps' decision and proceed with your project, you could challenge the permit denial itself in U.S. District Court. But in order to convince the court that the denial should be vacated and the Corps should review the matter again, you will need to show that the Corps acted arbitrarily and capriciously or otherwise not in accordance with law. Courts will be deferential to the agency's actions.

    If, on the other hand, you think the permit denial is a regulatory taking and you want money damages—just compensation—then you must bring the case in the U.S. Court of Federal Claims. The U.S. Court of Federal Claims has exclusive jurisdiction for takings cases involving more than $10,000. You may elect to challenge the permit denial in U.S. District Court and, at the same time, seek just compensation in the U.S. Court of Federal Claims. But each suit must seek different relief: it is injunctive relief in the U.S. District Court (an order to vacate the permit denial), and it is money damages in the U.S. Court of Federal Claims (the payment of just compensation). But if you try to proceed with both actions at the same time, the U.S. Court of Federal Claims will likely stay its case pending the resolution of the challenge to the permit denial. Why? Because if you prevail in U.S. District Court, then the matter will be remanded to the Corps, which might ultimately grant the permit. If the Corps grants the permit, your takings claim is rendered moot.

The *Penn Central* factors

If you are able to clear the procedural hurdles and the U.S. Court of Federal Claims considers your regulatory takings claim on the merits, it will conduct an ad hoc factual inquiry. There is no single, established formula for determining whether a permit denial constitutes a taking for which just compensation is due. The decision, as courts like to emphasize, will turn on the particular circumstances of each case.

The U.S. Supreme Court has, however, articulated several factors that are relevant to this ad hoc factual inquiry. In *Penn Central Transportation Company v. New York City*, the Court examined whether a landmark preservation law that prevented a 55-story tower from being placed on top of the Grand Central Terminal was a taking. The decision turned on three factors: the economic impact of the regulation on the property owner; the extent to which the regulation interferes with distinct investment-backed expectations; and the character of the government's action. Applying these factors to the denial of the permit for the tower, the Court held that no taking had occurred. The landmark law allowed the property to continue to be used as it had been and thus did not interfere with its present use or prevent a reasonable return on the property owner's investment. The Court also observed that the air rights over Grand Central Station could be transferred (and sold) to other parcels, thereby mitigating any economic impact caused by the permit denial.

In the context of wetland regulation, it is the first two *Penn Central* factors—the economic impact of the permit denial and the extent to which the permit denial interferes with reasonable investment-backed expectations—that are most relevant.

Applying the *Penn Central* factors: The *Florida Rock* saga

The *Penn Central* factors may seem relatively straightforward. To determine the economic impact of a permit denial, a court will compare the value of the property pre-denial and post-denial, along with what economically viable uses of the property remain. For reasonable investment-backed expectations, a court will inquire as to when the property was acquired and when investments were made. The case of Florida Rock illustrates that applying the *Penn Central* factors is not so simple. It also explains why the Corps is so concerned about takings cases.

In 1972, shortly before the Clean Water Act's enactment, Florida Rock bought a 1,560-acre parcel of land in Dade County for approximately $3 million (or $1,900 per acre). Florida Rock then applied for a section 404 permit to mine limestone beneath 98 acres of wetlands. The Corps denied the permit, and Florida Rock filed a takings claim in 1982. The case spanned many distinguished legal careers.

In 1990, the trial court concluded that the permit denial was a taking because it reduced the parcel's value 95 percent (from $10,000 per acre to a nominal value of $500 per acre). In reaching this conclusion, the trial court disregarded a market made up of speculators, reasoning that no one with full knowledge of the regulatory regime would be willing to gamble in purchasing the property. On appeal, the Federal Circuit vacated the lower court's decision and remanded the case.

The Federal Circuit stated that a relevant market for determining the present value of a parcel may be made up of investors who are speculating in whole or in part. A speculator, the Federal Circuit noted, may not be too concerned about present regulatory restrictions if it intends to hold the property for a number of years. The Federal Circuit observed that the public perception of the Everglades has shifted from yesterday's "mosquito haven" to "today's wetland to be preserved" and no one knows what tomorrow's view will bring, or how the market will respond. Accordingly, it was an error for the trial court to disregard this market of speculators.

On remand in 1999, the Court of Federal Claims determined that the post-denial value was $2,822 per acre, a 73.1 percent diminution in value. The court also found that the permit denial had interfered with Florida Rock's reasonable investment-backed expectations. Florida Rock had purchased the site before the Clean Water Act's enactment (1972) and had begun mining operations such as conducting a feasibility study, filling for a road, and stripping overburden before the Corps' assertion of jurisdiction over isolated wetlands (1977). Thus, the severe economic impact, coupled with the interference with reasonable investment-backed expectations, constituted a taking. Florida Rock was awarded $752,444, plus compound interest since 1982, and attorney fees.

In 2000, the Court of Federal Claims added one more twist to this ongoing saga. Much earlier in the litigation, the court held that only 98 out of the 1,560 acres were at issue. In this order, however, the court called this an "exceptional case," reversed itself, and declared that the takings claim regarding the 1,462 acres was ripe for review. This issue could have been the subject of an immediate appeal, which the court fervently hoped would "materially advance the ultimate termination of this litigation." The hope was well placed, as the parties settled in September 2001 for $21 million.

Of rats, rabbits, and reasonable investment-backed expectations

While *Florida Rock* was a victory for private property rights advocates, another decision arising out of Florida at the same time underscored the ad hoc nature of takings cases. In *Good v. United States*, a Philadelphia attorney familiar with wetland regulations purchased a 40-acre site, Sugarloaf Shores, located in Monroe County, Florida, in 1973.[6] Sugarloaf Shores contained 32 acres of wetlands. In the 1980s, Mr. Good obtained several Corps of Engineers permits to develop the site, including one in 1988, but was unable to secure authorization from the county and water management district.

In 1990, Mr. Good submitted a new plan to the Corps. Since his last application, however, the Lower Keys marsh rabbit had been listed as endangered under the Endangered Species Act (ESA). The Corps then entered into consultation with the Fish and Wildlife Service (FWS) as required under ESA section 7. Eventually, the Corps denied the 1990 application because of concerns about the endangered marsh rabbit (and the silver rice rat, which also had been declared to be endangered). Moreover, based on new information, the FWS found that the development authorized by the 1988 permit would jeopardize these species, but offered reasonable and prudent alternatives (RPAs) to mitigate the impacts. Mr. Good rejected the RPAs as economically infeasible, and the 1988 permit expired without any development occurring. He then filed a takings claim.

In a comprehensive opinion, the Court of Federal Claims granted summary judgment for the United States. The court rejected Mr. Good's contentions that the ESA required him to leave his property in its natural state, thereby destroying its economic value. The court observed that the ESA did not necessarily preclude all development of the site, noting that the FWS had recommended RPAs that would allow some development. Furthermore, the court noted that Sugarloaf Shores retained some value, even if it remained undeveloped. The fact that Mr. Good had signed a contract with an environmental consultant, which assigned the undeveloped site a value of $350,000, served to undercut his position.

In applying the *Penn Central* factors, the court focused on reasonable investment-backed expectations. Emphasizing that "property owners operating in a highly regulated field could not have a reasonable expectation that government regulation would not be altered to their detriment," the court examined the regulatory regimes in place at the time of acquisition and at the time of investment in development. When Mr. Good bought the site in 1973, the Corps had expanded its Rivers and Harbors Act program to encompass environmental concerns and the state had an existing comprehensive land use scheme. In 1980, when he first began to expend significant funds to develop the property, the Corps had already been asserting jurisdiction over wetlands under its Clean Water Act section 404 authority, and the ESA had been enacted. At all relevant times, Mr. Good knew that this environmentally sensitive property would be difficult to develop because of existing federal and state regulatory regimes. The court found no interference with reasonable investment-backed expectations and thus held that no taking had occurred.

Mr. Good appealed to the Federal Circuit, which unanimously affirmed the lower court's ruling, relying on the lack of reasonable investment-backed expectations. The Federal Circuit noted that the ESA did not exist when Mr. Good acquired the site, but stressed that he was aware of "the difficult regulatory path ahead" and that "rising environmental awareness translated into ever-tightening land use regulations." Yet, the Federal Circuit noted, he took no steps to obtain regulatory approvals for seven years. Because Mr. Good "must have been aware that standards could change to his detriment," the appellate court held that he had no reasonable investment-backed expectations.

Lucas v. South Carolina Coastal Council: No need to balance factors

In some cases, the ad hoc balancing approach of *Penn Central* is not used. As the U.S. Supreme Court explained in the 1992 case of *Lucas v. South Carolina Coastal Council*, a balancing approach is not needed when a regulation destroys all economically beneficial or productive use of land. While not a wetland case (and not involving federal law), the principles enunciated in *Lucas* are generally applicable to all regulatory takings cases. It was another victory for private property rights advocates that sent shock waves through regulatory agencies nationwide—at least at first. Its actual impact has been rather muted.

Mr. Lucas purchased two beachfront lots in 1986 for approximately $1 million. In 1988, the state of South Carolina passed the Beachfront Management Act, which flatly prohibited construction of occupiable improvements within a certain erosion zone. Mr. Lucas's two lots fell within that zone; the act permitted no exceptions to the prohibition.

He then filed a takings claim in the South Carolina Court of Common Pleas. The trial court found that the property had been rendered "valueless" and required just compensation of $1.2 million. The Supreme Court of South Carolina reversed, concluding that when regulation of the use of property is designed to prevent serious public harm, no just compensation is due regardless of the regulation's effect on the property's value. The U.S. Supreme Court then reversed the state court.[7]

Although a takings analysis generally requires an ad hoc, factual inquiry, the Court identified two discrete categories of governmental actions that are takings per se. The first is when an owner suffers a "physical 'invasion'" of his property. The second occurs when governmental action "denies all economically beneficial or productive use of land." The Court stated that "when the owner of real property has been called to sacrifice *all* economically beneficial uses in the name of the common good, that is, to leave his property economically idle, he has suffered a taking."

After articulating this "bright line" test, the Court then carved out a narrow exception. When acting to proscribe a use that is not inherent in the property's title, the government need not provide just compensation, even if the regulation renders the property valueless. The Court limited this exception to "the restrictions that background principles of the State's law of property and nuisance already place upon land ownership." Accordingly, prohibition of common law nuisances (or enforcement of a preexisting public trust doctrine or navigable servitude) does not result in a taking. The Court provided an example relevant to wetland regulation, stating that the government need not compensate an owner who is denied a permit "to engage in a landfilling operation that would have the effect of flooding others' land."

On remand, the state court found that building a coastal home was not a nuisance. The Coastal Council eventually paid Mr. Lucas approximately $1.5 million for the two lots, about $500,000 of which went to attorney fees. Regulatory agencies, including the Corps and the EPA, shuddered.

But they did not need to worry. The impact of *Lucas*, at least as measured in lost regulatory takings cases, would be limited.

The irrelevance of *Lucas*

The categorical taking rule of *Lucas* is only applicable when the property at issue has been rendered valueless. When examining how a permit denial has affected a property, the court must decide whether to look at only the restricted portion of the property (e.g., the wetlands) or the "parcel as a whole." Courts will typically consider the parcel as whole. If the parcel as a whole (wetlands and uplands) is at issue, then the property retains some value. The uplands can at least be put to some use, and thus *Lucas* is not applicable.

The case of *Forest Properties, Inc. v. United States* illustrates the importance of the "parcel as a whole" determination. FPI intended to develop property contiguous to Big Bear Lake in San Bernardino County, California. The project site was 62 acres: 57.6 acres of upland and 4.4 acres of the lakebottom.

The Corps denied a section 404 permit to fill the lake. FPI abandoned the project's lakebottom portion and proceeded to develop the uplands. FPI also proceeded to the Court of Federal Claims, asserting that the Corps' denial was a taking.

The court ruled that the permit denial did not constitute a taking. The dispositive factor was the determination of the relevant parcel. Rather than focusing on the lakebottom property alone (which probably would have required a *Lucas* analysis), the court found the relevant parcel to be the entire project site—lakebottom and uplands. The court noted that the areas were contiguous, that they were acquired only five months apart, and that they were intended to be used as one income-producing unit. Because the relevant parcel retained substantial value even after the permit denial, the court then applied the *Penn Central* factors and ruled against FPI.

Even if a court focuses solely on the burdened portion of the property (or if the parcel as a whole is entirely wetlands), *Lucas* may still be inapplicable because of the growth of mitigation banking. Under a mitigation banking system, a wetland can have an economic value by remaining a wetland (Gardner, 1996). A landowner can restore or enhance low-value wetlands or preserve high-value wetlands to produce wetland credits that can be sold to permittees. By providing economic value to wetlands in their natural state, mitigation banking can effectively consign *Lucas* to law school casebooks. It is an interesting decision, but it has little continuing resonance in takings litigation.

Reasonable investment-backed expectations revisited

A U.S. Supreme Court case that does still matter is *Palazzolo v. Rhode Island*, where the Court reemphasized that the *Penn Central* approach requires a balancing of factors. Located on Rhode Island's Atlantic coast, the property at issue contained 18 acres of tidal marshes and a few upland acres. Shore Gardens, Inc. (SGI) obtained the property in 1959; Mr. Palazzolo was SGI's president and became its sole shareholder in 1960. The state eventually denied a 1963 application to erect a bulkhead and fill the site. A 1966 application to fill the site to construct a beach club met a similar fate.

In 1977, the Coastal Resources Management Council (CRMC), the state agency with regulatory jurisdiction over coastal wetlands, promulgated regulations that prohibited the filling of such areas without a special exception. In 1978, SGI's corporate charter was revoked, and Mr. Palazzolo became the property's owner by operation of law (figure 11-1).

Mr. Palazzolo sought permission in 1983 to erect a bulkhead (similar to the 1963 proposal); the CRMC denied the permit. In 1985, he sought

FIGURE 11-1. Mr. Palazzolo on his property. (Photo credit: R. Gardner.)

permission to build a beach facility (similar to the 1966 proposal); this too was denied.

Mr. Palazzolo filed a takings claim in Rhode Island state court. He contended that he lost $3.15 million in profits because he was unable to proceed with single-family homes on the property's 74 lots. The Superior Court found for CRMC, and the Rhode Island Supreme Court affirmed.

The court decided that the government's actions did not constitute a categorical taking under *Lucas*: the land still retained significant value (the upland portion, if developed, was valued at $200,000, while the wetland portion was valued at $157,500). Turning to the *Penn Central* test, the court pointed out that regulations restricting the filling of coastal wetlands were in effect when Mr. Palazzolo acquired the property. Accordingly, any permit denial did not frustrate reasonable, investment-backed expectations. In the court's view, the timing of the acquisition was dispositive.

The U.S. Supreme Court rejected the notion that the timing of Mr. Palazzolo's acquisition automatically barred his takings claim: "A blanket rule that purchasers with notice [of land use regulations] have no compensation right when a claim becomes ripe is too blunt an instrument to accord with the duty to compensate for what is taken." In her concurrence, Justice O'Connor (who supplied the fifth vote on this point) noted that the timing of acquisition relative to the regulatory scheme's enactment, while not dispositive, remains relevant to the issue of the reasonableness of investment-backed expectations.

The Court agreed with the Rhode Island Supreme Court that Mr. Palazzolo did not present a *Lucas* taking because the parcel's upland portion retained some value. Accordingly, the case was remanded to be considered under the *Penn Central* factors, in light of the Court's discussion of investment-backed expectations.

Although victorious in the U.S. Supreme Court, Mr. Palazzolo did not fare well on remand. In July 2005, a Rhode Island Superior Court rejected his takings claim in part based on Rhode Island's public nuisance law and public trust doctrine. The court stated that "[d]espite wishful thinking on Palazzolo's part, he paid a modest sum to invest in a proposed subdivision that he must have known from the outset was problematic at best. . . . Constitutional law does not require the state to guarantee a bad investment."

The most sympathetic takings plaintiff

Who then would make the most sympathetic takings plaintiff? The few successful wetland takings claims share two characteristics. First, the plaintiff suffered a total or a nearly total wipeout of the economic value of the property as a result of a permit denial. Second, the plaintiff had typically purchased the property before it (or the proposed activity) had become subject to regulation. (It also helped if Judge Loren Smith heard your case.[8]) This limited pool of potential plaintiffs, however, is not likely to grow, unless Congress amends the Clean Water Act to (re)assert jurisdiction over isolated wetlands. (And Judge Smith has retired.)

But bringing a takings case is not for the fainthearted. As the twenty-year saga of *Florida Rock* demonstrates, a takings case requires time and money. Even then, the government has a range of defenses—from procedural hurdles to statute of limitations to pointing out alternative uses of the property—that can defeat a takings claim. Accordingly, disappointed property owners should probably not proceed down this path unless they have a certain ideological fervor or are looking for a new hobby.

How should the Corps weigh the risks of a takings case?

When making a section 404 permit decision, the Corps should not factor in the risks associated with a takings claim for several reasons. First, as a practical matter, the bark of the takings claim is worse than the bite; the government rarely loses. (The counterargument might be that the government rarely loses because the Corps denies so few permits.) Second, as a legal matter, the Corps is obligated to apply the section 404(b)(1) guidelines to a proposed action, and these regulations focus on the environmental impacts of a project. The section 404(b)(1) guidelines do not provide the authority to issue a permit out of fear of takings litigation (whether warranted or not). A permit should be granted or denied based on its environmental impact.

When a court concludes that the Corps' denial of a permit is in fact a taking, this does not mean that the Corps did anything wrong. If the Corps had not followed the applicable statute and regulations, then its decision could be vacated as arbitrary, capricious, or not otherwise in accordance with law. But a takings action presumes that the Corps acted lawfully in following the relevant regulations. The claim is that by following these rules, the property owner has been deprived of its property. And, as we noted at the outset, the government can take your property, it just has to pay you for it.

Nevertheless, the Corps does try to be sensitive to impacts on private property rights. Furthermore, the Corps is particularly sensitive to congressional inquiries on behalf of constituents, and this political pressure can persuade the Corps to exercise its discretion by granting permits that should have been subjected to greater conditions or denied outright. In such cases, the government is spared a takings claim, but the public still pays. Only this time, the payment is in lost ecosystem services, the public benefits that wetlands provide.

Chapter 12

Concluding Thoughts and Recommendations

The BP/Deepwater Horizon oil spill is the biggest single-incident disaster for the aquatic environment in U.S. history. It is estimated that more than 200 million gallons of crude oil were released into the Gulf of Mexico, soiling shores from Texas to Florida. The coastal wetlands of Louisiana have been particularly hard-hit. Billions of dollars will be spent on litigation, remediation, and compensation, but that money will not offset the damage to the environment and the ecosystem services lost.

This human-made tragedy sadly underscores the importance of understanding administrative law and influences on agency behavior. For example, in 2000 the Minerals Management Service (MMS), an agency within the U.S. Department of Interior that regulated offshore oil and gas activities, contemplated requiring oil rigs to use an acoustical regulator (Gold et al., 2010). We will never know if the device, which is required in Norway and Brazil, would have averted the catastrophe in the Gulf of Mexico because in 2003 MMS decided against requiring these secondary backups.[1]

Of course, the primary blame (and liability) for the oil spill must rest with the corporations and individuals who were responsible for actions on the Deepwater Horizon. But the disaster in the Gulf also illustrates a worst-case scenario of what happens when an agency fails to regulate in a comprehensive and diligent fashion. With that wound fresh in mind, let us consider several recommendations to improve efforts to protect wetlands in the United States.

- **Congress should pass the Clean Water Restoration Act to remove doubts about federal jurisdiction over wetlands by deleting the reference to "navigable waters" in the Clean Water Act.** The EPA and the Corps have attempted to interpret what the terms "navigable waters" and "waters of the United States" mean, but the U.S. Supreme Court, to be polite, has made a hash of it. Instead of relying on case-by-case court decisions or a lengthy agency rulemaking, Congress should simply settle the debate. The cumulative impacts analysis that was the basis of *Gonzales v. Raich* provides support that such a law is within Congress's Commerce Clause powers and would pass constitutional muster.[2]

- **The Corps should reorient its regulatory philosophy.** Often the project proponent—the permit applicant—is viewed as the customer. Indeed, Corps districts have customer satisfaction survey forms for permit applicants (Connolly, 2007). Although the Corps should strive to make its procedures "user friendly," the Corps should remember that the public is the ultimate customer. The Corps' primary obligation ought to be the goals of the Clean Water Act: to restore and maintain the chemical, physical, and biological integrity of the nation's waters. The primary obligation should not be to process permit applications as quickly as possible; it should be to protect the ecosystem services provided to people by wetlands.

- **The Corps needs to emphasize avoidance of wetland impacts and deny more permits.** As noted by Wilkinson et al. (2009), while much of the Corps' and the EPA's focus has been on "improving the third step in the mitigation process—compensatory mitigation," the agencies have paid little attention to developing "tools, guidance and/or regulations to ensure the rigorous application of [the] avoidance" requirement. Avoidance especially ought to be invoked with respect to wetlands that are difficult to restore (NRC, 2001). Furthermore, these permit decisions should be based on ecological factors, not concerns about regulatory takings. Under current case law, it is very unlikely that a permit denial will constitute a regulatory taking. Moreover, the development of a mitigation banking regime whereby wetlands have an economic value as a mitigation site further insulates the government from having to pay just compensation.

- **The EPA should maintain its veto authority and not hesitate to unsheathe it.** For many years, I thought that the EPA's veto authority was a withered appendage that provided no benefits and that therefore should be amputated. The veto of the Yazoo Backwater

Area Project, which saved 67,000 acres of bottomland hardwoods, convinced me otherwise. If the Corps declines to deny more permits, the EPA should exercise its authority under Clean Water Act section 404(c) to veto projects that "will have an unacceptable adverse effect on municipal water supplies, shellfish beds and fishery areas (including spawning and breeding areas), wildlife, or recreational areas."

- **The Corps should eliminate NWP 21.** Nationwide permits, like all general permits, are only permissible for a category of activities if the activities will result in minimal adverse environmental impacts on an individual and a cumulative basis. An NWP for surface coal mining operations is oxymoronic. The cumulative impacts alone from coal mining should preclude the use of a general permit (Palmer et al., 2010). Such activities should proceed through the individual permit process. In June 2010 the Corps announced that it was suspending the use of NWP 21 in the Appalachian region. This is a very positive development, and the Corps should take the next step to eliminate NWP 21 altogether. If the Corps declines to take such action, the states could force the same result by withholding state water quality certifications.

- **The Corps should make permit decisions on a watershed basis, taking into account cumulative impacts.** This recommendation echoes one made by the National Research Council in 2001. The 2008 Corps and EPA regulation on compensatory mitigation requires the Corps to employ a watershed approach, which is a great opportunity for the Corps to move from reactive permit review to proactive planning. A good watershed plan can be an effective tool to help guide avoidance, minimization, and compensation, as well as decision making based on a cumulative impact analysis. However, moving the approach from concept to reality will require additional resources, training, and guidance.

- **The Corps and the EPA should identify minimization best practices.** In the avoid-minimize-compensate sequence, the debate typically focuses on avoidance and compensation. The middle step is sometimes overlooked (Wilkinson et al., 2009). The Environmental Law Institute (2008) has published a compilation of minimization requirements in state programs, which is a good starting point.

- **The Corps should implement the regulatory preference for compensation from mitigation banks; at the same time, the Corps should tighten up on early credit releases.** The 2008

Corps and EPA regulation establishes a compensatory mitigation hierarchy that looks to mitigation banks first. But the regulation's preference can be viewed as soft or hard, depending on the reader's perspective. In either case, Corps districts will have discretion on how to apply it in the field. Based on the shaky track record of permittee-responsible mitigation, I recommend a strong preference for mitigation banks, as long as early releases are limited. A significant percentage of a bank's total credits should not be released until it has met all of its performance standards and has identified a long-term steward of the site. Otherwise, one of the justifications for mitigation banking's top spot in the mitigation hierarchy (timing) is undercut.

- **Compensatory mitigation sites must be monitored by the agencies.** This recommendation also harkens back to the 2001 National Research Council report, which called for effective monitoring of compensatory mitigation sites. The Corps and the EPA should actually visit mitigation sites, including mitigation banks, to assess whether they are meeting their performance standards. To encourage this, the Corps' personnel performance standards must be adjusted to put greater emphasis on time spent in the field rather than permit approval. The Corps and the EPA also should consider partnerships with local colleges and universities to conduct long-term studies on the performance of mitigation sites (Gardner et al., 2009).

- **The Corps and other agencies should focus on the long-term stewardship of compensatory mitigation sites.** There is no point to all the time, energy, and money spent on compensatory mitigation if the agencies do not ensure that each site has a long-term management plan. A conservation easement or deed restriction is not sufficient. Each compensatory mitigation site should have a responsible party that is dedicated to protection of the site and that has the funds necessary to do so. The 2008 regulation gives Corps districts the authority to require endowment accounts for all types of mitigation—banks, in-lieu fee programs, and permittee-responsible mitigation. Let's see how that authority is really applied; permittee-responsible mitigation should not be held to a less exacting standard.

- **The Corps should encourage and accept preservation of high-quality wetlands as compensatory mitigation, including "preservation only" packages.** The preamble to the 2008 regulation observes that "[p]reservation is rarely the sole source of compensatory mitigation" for a section 404 permit. Yet a basic conservation approach is to protect biodiversity hotspots, especially large, contiguous blocks of habitat that provide migration corridors. The importance of protecting existing natural resources is recognized at the international level. In crafting post-Kyoto strategies such as REDD (reduction in emissions from deforestation and forest degradation), international negotiators in the climate change debate acknowledge that developing countries need incentives to protect existing forests, including peat swamps (UN-REDD Programme, 2010). We should have similar incentives domestically to protect existing high-quality wetlands within the Clean Water Act section 404 program and encourage preservation to be used more frequently to offset impacts to low-quality wetlands. The ratio should not be 1:1, however. The Nature Conservancy's Virginia Aquatic Resources Restoration Trust Fund (discussed in chapter 8) is currently providing a 16:1 ratio. Any ratio in that neighborhood (in the double digits) should override concerns about preservation's contribution to no net loss. But if objections remain that preservation alone does not contribute to the goal of no net loss, then preservation should be encouraged as part of a mitigation package that includes restoration of other areas.

- **The federal government should maintain and create incentives for wetland restoration, including the possibility of ecosystem credit stacking.** Congress should continue to fund voluntary restoration efforts such as the Wetlands Reserve Program, Conservation Reserve Program, and Environmental Quality Incentives Program and expand opportunities for perpetual conservation easements. Congress and the agencies also should examine appropriate opportunities to allow credit stacking, a more controversial proposition. Credit stacking generally refers to establishing two or more different types of ecosystem credits on the same parcel of property (Fox, 2008). For example, one spatially overlapping area might produce wetland, endangered species, water quality, and/or carbon sequestration credits. Credit stacking would permit the owner of the credits to sell them in different markets and thus have a diversified revenue stream. One environmental benefit of stacking is that it may provide a financial incentive for some property owners to conserve their land when they might otherwise be reluctant to do so. A danger is that a property owner might be essentially selling the same environmental benefit multiple times. Firm policies have yet to be set on credit stacking; it is a contentious issue that represents the next great mitigation debate, especially if protocols are developed to measure the extent to which certain types of wetlands sequester greenhouse gases (Emmett-Mattox et al., 2010).[3]

- **Congress should discontinue perverse incentives that contribute to wetland destruction.** One effect of the federal government subsidizing ethanol production is to encourage farmers to remove land enrolled in the Conservation Reserve Program and similar conservation initiatives. For example, the Prairie Pothole Region and its associated prairie-wetland habitat are at risk from the increase in corn planting (Brooke et al., 2009). As these areas are converted to cropland, critical habitat for migratory waterfowl will be lost. To add (economic) insult to (ecological) injury, ethanol subsidies rely on taxpayer dollars (Hassett, 2006). Environmental groups should continue to work with unlikely allies (e.g., organizations concerned about taxes) to phase out and end such subsidies. Realistically, however, probably nothing will be done to halt the pandering to the corn lobby until the Iowa caucuses no longer play a prominent role in presidential primaries.

- **The Corps and the EPA should increase enforcement efforts and use technology to do so.** Since 2003, the Massachusetts Department of Environmental Protection has used infrared aerial photography and a computer program to successfully identify illegal wetland fills throughout the state (Giles, 2005). The Corps and the EPA (and potential citizen suit plaintiffs) should avail themselves of this technology to ferret out wetland violators and increase enforcement actions.

- **The Corps should provide more transparency in the section 404 program.** In April 2010 a group of environmental organizations and academics (and the National Mitigation Banking Association) wrote to the Assistant Secretary of the Army (Civil Works) to urge greater transparency in the Corps' regulatory program. In particular, the letter requested that the Corps make available on the Internet data on jurisdictional determinations, permit information (including the location of authorized impacts), compensatory mitigation projects, monitoring and compliance reports, and records such as final decision documents (including mitigation plans). A host of stakeholders could use this information for independent evaluations of compensatory mitigation projects, economic planning, and watershed protection efforts. The increased transparency also would make the Corps more accountable to its ultimate customer, the public.[4]

To implement this wish list, we would need the support of many different players. Obviously, many of these actions would require substantial funds, and Congress would need to be willing to provide an appropriate level of appropriations to the Corps and the EPA. Executive branch officials would also need to modify policies and priorities (and the Corps SOP) to emphasize watershed planning, monitoring, enforcement, and transparency. But even then, the individual regulator in the field will still retain administrative discretion. It will be up to interested stakeholders to participate in the agency decision-making process—at all levels—to channel that discretion for the protection of wetlands.

Epilogue: Where Are They Now?

We have covered quite a bit of ground. Let us now look back at some of these wetland and aquatic sites, as well as a few individuals that we considered, to see what has happened to them as a result of our wetland laws, policies, and politics. There are a few surprises.

Chapter 1

EVERGLADES NATIONAL PARK (ENP)

In 2000, Congress approved a Comprehensive Everglades Restoration Plan (CERP). If completed, the CERP would be among the largest ecosystem restoration projects in history. Its objectives include restoring the health of ENP, as well as improving water supply and flood protection in South Florida. The CERP requires the participation of multiple federal, state, local, and tribal partners and contemplates the construction of reservoirs, removal of levees, and creation of stormwater treatment areas on former agricultural land. The original cost of the project was estimated to be $7.8 billion; by 2008 the figure was north of $12.5 billion.

In June 2008, Florida Governor Charlie Crist announced that the state of Florida would purchase 180,000 acres in the Everglades Agricultural Area for $1.75 billion from U.S. Sugar. The deal has since been reduced in scope and price and, not surprisingly, was the subject of litigation. Critics contend that the state is overpaying for the land, giving a sweetheart deal to a long-time polluter, and diverting funds from other CERP projects. Proponents of the deal argue that the purchase of the property (at least in the original deal) would obviate the need for CERP's complex system of reservoirs and wells, thereby allowing the restoration of the Everglades to be simpler and less expensive. Moreover, they note that if the state does not acquire the land, it may be sold off piecemeal to developers and others. In October 2010 Florida purchased approximately 27,000 acres for $197 million.

WETLANDS PRESERVE

Established in 1989, this live-music nightclub embraced environmental and social activism and education as a primary mission. Although the nightclub closed in 2001, Wetlands Preserve now operates as the Wetlands Activism Collective, describing itself as "a volunteer-run grassroots organization based in New York City, focused on resisting global capitalism and its devastating effect on the environment and the lives of human and nonhuman animals." The Wetlands Activism Collective is active with Global Justice for Animals and the Environment, which challenges free-trade policies. More information is available at http://wetlands-preserve.org.

Chapter 3

RIVERSIDE BAYVIEW HOMES

In 1986, the U.S. Supreme Court upheld the federal government's jurisdiction over wetlands adjacent to a traditional navigable water, halting a proposed housing project in Michigan. According to the Harrison Township Assessor's Office, the property was later used as off-site compensatory mitigation to offset permitted impacts elsewhere.

SOLID WASTE AGENCY OF NORTHERN COOK COUNTY (SWANCC)

SWANCC was searching for a site for a nonhazardous waste landfill. The Corps denied a section 404 permit in part because the site had a blue heron rookery. SWANCC's quest led to the U.S. Supreme Court, which invalidated the Migratory Bird Rule and called into question the federal government's jurisdiction over isolated waters. With no need for a section 404 permit, SWANCC could then proceed with the project. After the decision, however, the state of Illinois expressed an interest in acquiring the property. In December 2001, SWANCC sold most of the site to the state for $21 million. The site is now the Heron Woods State Habitat Area and is managed by the Illinois Department of Natural Resources.

JOHN RAPANOS

Rapanos was convicted of filling wetlands without a permit and faced civil and criminal penalties. In a fractured ruling that brought us the "significant nexus" test, the U.S. Supreme Court vacated his conviction and remanded the case back to the lower courts for further proceedings. Rapanos eventually settled his dispute with the federal government, but at a heavy price. To resolve the civil enforcement action, he paid a $150,000 civil penalty and agreed to restore the 54 acres of wetlands illegally filled, which will cost an estimated $750,000. In addition, Rapanos agreed to preserve another 134 acres of wetlands whereby the state of Michigan held a conservation easement. To settle the criminal action, Rapanos agreed to accept the conviction (and stop all appeals) in exchange for no prison time. He also had to pay a criminal fine of $185,000.

Rapanos's son is reportedly starting a wetland mitigation bank in Michigan.

Chapter 4

COEUR ALASKA

Coeur Alaska wanted to reopen the Kensington gold mine in southeastern Alaska and discharge the slurry into Lower Slate Lake. In 2009, the U.S. Supreme Court held that slurry constituted fill material and thus the Corps was the proper permitting agency. The Kensington gold mine opened ahead of schedule in June 2010. Coeur d'Alene Mines Corporation's stock price jumped 7 percent in one day in May 2010 as gold prices hit an all-time high.

Chapter 5

ZABEL V. TABB

In 1967, the Corps denied—for the first time—a Rivers and Harbors Act section 10 permit on environmental grounds when it prevented a developer

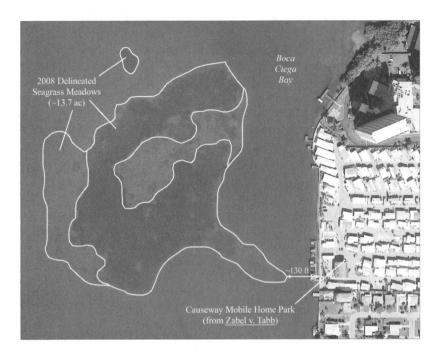

FIGURE 13-1. Boca Ciega Bay. (Source: adapted from the South Florida Water Management District.)

from filling 11 acres of Boca Ciega Bay in Pinellas County, Florida, for a trailer park. That portion of the bay remains unfilled today, with a seagrass meadow where part of the trailer park would have encroached (figure 13-1).

WALTON TRACT

In *Fund for Animals v. Rice*, environmental groups sued the Corps for issuing a permit to Sarasota County, Florida, for a landfill. The Eleventh Circuit Court of Appeals upheld the permit, and Sarasota proceeded with the construction of the Central County Solid Waste Disposal Complex. A permit condition required the county to restore 126.6 acres of slough wetlands and enhance another 136.3 acres of slough wetlands. The Corps deemed the compensatory mitigation successful. The county, through its Natural Resources Department, has continued to manage the site, removing invasive plants and animals such as hogs. The Solid Waste Department funds these efforts. The area is now called the Pinelands Reserve.

OLD CUTLER BAY

As part of a permit elevation in 1990, Corps headquarters found that the project description with its specific reference to a Jack Nicklaus signature championship golf course inappropriately truncated the section 404(b)(1) alternatives analysis. Accordingly, Corps headquarters instructed the Jacksonville District to reevaluate the permit decision. In June 1996, the Jacksonville District issued a section 404 permit for a residential development to Old Cutler Bay Estates. The approved design spared most of the high-quality mangroves that had been at issue in the elevation. The permittee dropped the request for a golf course, reportedly deciding that it would be more profitable to construct additional homes on that portion of the site. The project, however, was not immediately built. Instead, the property (and permit authorization) was transferred several times. Shoma Homes eventually constructed a subdivision on the site (without a golf course) called Cutler Cay in 2007.

SWEEDENS SWAMP

In 1986, the EPA vetoed a Corps permit that authorized the filling of Sweedens Swamp in Attleboro, Massachusetts, for a shopping mall. The EPA had determined that the red maple swamp provided excellent wildlife habitat and that its destruction would result in unacceptable adverse impacts to wildlife and habitat. Today the site remains undeveloped and is largely privately owned. In a 2008 report, the Southeastern Regional Planning and Economic Development District described the site's condition as "a large red maple swamp complex interspersed with open marsh/scrub shrub habitat. Several potential vernal pools have been mapped in the area. No rare species have been identified in this area, although Sweedens Swamp provides an important natural habitat within an otherwise urban landscape." The report identified it as a priority protection area.

Chapter 6

LOUISIANA COASTAL WETLANDS

Invasive species such as nutria are currently the least of this system's challenges. In 2005, Hurricanes Katrina and Rita converted about 217 square miles—about 139,000 acres—of coastal marshes in Louisiana to open water. Recall that in the previous six years, the FWS reported that the United

States averaged a net gain of 32,000 acres of wetland per year. The losses in the immediate aftermath of the hurricanes wiped out more than four years of national wetland area gains. As of 2008, the U.S. Geological Survey reported that only approximately 12,000 acres of land (19 square miles) had returned. The more recent disaster is the 2010 BP/Deepwater Horizon oil spill. The extent of the damage from the spill is uncertain at this time. What is certain, however, is that the costs associated with any remediation, along with the value of the ecosystem services lost, will be billions of dollars.

Chapter 7

DISNEY WILDERNESS PRESERVE

Disney obtained the 8,500-acre Walker Ranch and established the Disney Wilderness Preserve in 1992 as off-site mitigation. Through an agreement with the Greater Orlando Aviation Authority, the Disney Wilderness Preserve was expanded by 3,000 additional acres. A Conservation Learning Center was constructed at the preserve in 1998. The Nature Conservancy now owns and manages the site, and it reports that the preserve is "home to more than 750 native plant and 300 animal species." It is open to the public, and I can personally attest that a visit and a walk on its trails are a welcome respite from the sensory overload of the nearby theme parks.

PEMBROKE PINES

Florida Wetlandsbank was an early entrepreneurial wetland mitigation bank that restored sawgrass marsh, wet prairie, and cypress flats on city-owned property in Pembroke Pines, Florida. The 450-acre site, now called Chapel Trail Park Nature Preserve, is managed and maintained by the city. The site offers visitors a boardwalk from which to observe wildlife, a picnic area, canoe rentals, and an environmental interpretative center for school groups. A 2005 assessment of mitigation banks in Florida highlights the distinction between legal compliance and ecological outcomes. The report found that although the bank had been deemed to meet its permit success criteria, the site had not achieved full wetland functions when compared to reference wetland conditions. The report noted, however, that when a bank is placed in a highly developed area, "the location and associated landscape support will always be a limiting factor in achieving full wetland function" (Reiss et al., 2007).

PANTHER ISLAND MITIGATION BANK

The 2,778-acre mitigation bank site was transferred to the Audubon Society, which incorporated it into an expanded Corkscrew Swamp Sanctuary. In 2009, the United States, through the FWS, designated Corkscrew Swamp Sanctuary as a Wetland of International Importance under the Ramsar Convention, a wetland conservation treaty. Corkscrew was the twenty-sixth Ramsar site in the United States, joining other wetland jewels such as Everglades National Park and Okefenokee National Wildlife Refuge. Ramsar Secretary General Anada Tiega traveled from the Ramsar Secretariat in Switzerland to attend a designation ceremony in February 2010 (figure 13-2).

FIGURE 13-2. Ramsar designation ceremony at Corkscrew Swamp Sanctuary. (Photo credit: Nancy DeNike. Reproduced with the permission of the Wetlandsbank Group)

MUD SLOUGH WETLAND MITIGATION BANK

One of the early banks in Oregon, it is recognized as an Important Bird Area by the Portland Audubon Society. In April 2010, the Corps' Portland District issued a public notice for a fourth phase to add 47 acres to the bank, which would involve "converting uplands and farmed wetlands to wet prairie and emergent wetlands dominated by native plants." This would increase the total area of the bank to about 233 acres. Although in the initial phases of the bank the Corps did not require an endowment account for long-term management, the landowners nevertheless voluntarily established long-term care provisions, and the Corps will require a funded endowment before the third and fourth phases of the bank are closed out.

MONASTERY OF THE HOLY GHOST

The Trappist monks in Conyers, Georgia, added stream credits to their bank in 2005 and also established the Honey Creek Mitigation Bank. The monastery's Web site explains that "[t]hese decisions, made after prayer and prudent discussion, allow us to be better stewards of the land as we look toward ecological restoration, conservation, and protection of the land for the good of all." Information about this and other banks in the Corps' Savannah District is now publicly available through the Regional Internet Bank Information Tracking System (RIBITS) on the district's Web site.

Chapter 8

VIRGINIA AQUATIC RESOURCES RESTORATION TRUST FUND

The Nature Conservancy of Virginia (TNC-Virginia) has run this in-lieu fee program since 1995, offsetting 241 acres of wetland impacts with more than 630 restored acres and 18,400 preserved acres. In the 2008 compensatory mitigation regulation, the Corps and the EPA tightened up the rules governing the use of in-lieu fee programs. For example, in-lieu fee programs now have to develop a watershed approach to selecting mitigation sites. Existing programs such as the Virginia Aquatic Resources Restoration Trust Fund were given until June 2010 to make any revisions necessary to comply with the new regulation. Corps districts could grant an additional three-year extension. TNC-Virginia's draft in-lieu fee instrument went out for public comment in December 2009 and, as of December 2010, is still being reviewed by the Corps.

Chapter 10

UNITED STATES V. CUNDIFF

The U.S. District Court found that the Cundiffs had violated the Clean Water Act for damaging 200 acres of wetlands. The court imposed $225,000 in civil penalties, all but $25,000 of which would be suspended if the defendants completed a restoration plan. The defendants appealed, claiming that the wetlands were not "waters of the United States" under *Rapanos*. In February 2009, the Sixth Circuit Court of Appeals concluded that the site met both Justice Kennedy's "significant nexus" test and the plurality's approach; accordingly, the penalties were upheld. The U.S. Supreme Court declined to review the case in October 2009.

JOHN POZSGAI

Convicted of forty counts of Clean Water Act violations for filling wetlands without a permit, he was sentenced to serve twenty-seven months in prison and ordered to restore the site. Pozsgai spent about a year and a half in prison and then finished his sentence at a halfway house. He petitioned for a pardon from President George H.W. Bush, who declined the request. In March 2007, he was brought back into the U.S. District Court for the Eastern District of Pennsylvania for his continuing refusal to restore the site. Pozsgai argued that *Rapanos* had changed the law and that his wetland was not a water of the United States. The judge disagreed and held him in contempt. Pozsgai's appeal to the Third Circuit Court of Appeals was dismissed for his failure to file a brief in a timely manner.

BIG HILL ACRES

Robert J. Lucas and the other defendants appealed their Clean Water Act, mail fraud, and conspiracy convictions on the grounds that the wetlands on Big Hill Estates were not "waters of the United States" in light of *SWANCC* and *Rapanos*. In a February 2008 decision, the Fifth Circuit affirmed the convictions, holding that the evidence presented at trial established that the wetlands had a "significant nexus" to a traditional navigable water. The U.S. Supreme Court declined to review the case. According to the U.S. Bureau of Prisons Web site, Lucas is serving his time in Montgomery, Alabama, with a release date of November 27, 2015.

Chapter 11

LUCAS V. SOUTH CAROLINA COASTAL COUNCIL

After David Lucas acquired two beachfront lots in 1986, the state of South Carolina passed a law prohibiting the new construction of habitable dwellings within a coastal erosion line, where Lucas's two lots were located.

The U.S. Supreme Court found that the law had resulted in a regulatory taking, and South Carolina paid Lucas approximately $1.5 million. South Carolina was then the owner of the lots. During the litigation, South Carolina amended its law to allow for special permits. After the state paid Lucas, it turned around and sold the properties to a developer for $750,000. One response (Goldstein and Goldstein, 1993) summed up South Carolina's hypocritical maneuverings aptly: "it is nice to be an environmentalist when it is at someone else's expense."

Clean Water Act (excerpts)

Section 101: Congressional declaration of goals and policy

(a) The objective of this Act is to restore and maintain the chemical, physical, and biological integrity of the Nation's waters.

* * *

Section 404: Permits for dredged or fill material

(a) Discharge into navigable waters at specified disposal sites

The Secretary may issue permits, after notice and opportunity for public hearings for the discharge of dredged or fill material into the navigable waters at specified disposal sites.

* * *

(b) Specification for disposal sites

Subject to subsection (c) of this section, each such disposal site shall be specified for each such permit by the Secretary:

(1) through the application of guidelines developed by the Administrator, in conjunction with the Secretary, which guidelines shall be based upon criteria comparable to the criteria applicable to the territorial seas, the contiguous zone, and the ocean under section 403(c)

* * *

(c) Denial or restriction of use of defined areas as disposal sites

The Administrator is authorized to prohibit the specification (including the withdrawal of specification) of any defined area as a disposal site, and he is authorized to deny or restrict the use of any defined area for specification (including the withdrawal of specification) as a disposal site, whenever he determines, after notice and opportunity for public hearings, that the discharge of such materials into such area will have an unacceptable adverse effect on municipal water supplies, shellfish beds and fishery areas (including spawning and breeding areas), wildlife, or recreational areas.

* * *

(d) "Secretary" defined

The term "Secretary" as used in this section means the Secretary of the Army, acting through the Chief of Engineers.

(e) General permits on State, regional, or nationwide basis

(1) In carrying out his functions relating to the discharge of dredged or fill material under this section, the Secretary may, after notice of opportunity for public hearing, issue general permits on a State, regional, or nationwide basis for any category of activities involving discharges of dredged or fill material if the Secretary determines that the activities in such category are similar in nature, will cause only minimal adverse environmental effects when performed separately, and will have only minimal cumulative adverse effects on the environment.

* * *

Section 502: General Definitions

Except as otherwise specifically provided, when used in this chapter:

* * *

(6) The term "pollutant" means dredged spoil, solid waste, incinerator residue, sewage, garbage, sewage sludge, munitions, chemical wastes, biological materials, radioactive materials, heat, wrecked or discarded equipment, rock, sand, cellar dirt and industrial, municipal, and agricultural waste discharged into water.

* * *

(7) The term "navigable waters" means the waters of the United States, including the territorial seas.

* * *

(12) The term "discharge of a pollutant" and the term "discharge of pollutants" each means (A) any addition of any pollutant to navigable wa-

ters from any point source, (B) any addition of any pollutant to the waters of the contiguous zone or the ocean from any point source other than a vessel or other floating craft.

* * *

(14) The term "point source" means any discernible, confined and discrete conveyance, including but not limited to any pipe, ditch, channel, tunnel, conduit, well, discrete fissure, container, rolling stock, concentrated animal feeding operation, or vessel or other floating craft, from which pollutants are or may be discharged. This term does not include agricultural stormwater discharges and return flows from irrigated agriculture.

* * *

EPA Regulations 40 CFR Part 230 (excerpts)

Section 404(b)(1) Guidelines for Specification of Disposal Sites for Dredged or Fill Material

230.1 Purpose and policy.

* * *

(c) Fundamental to these Guidelines is the precept that dredged or fill material should not be discharged into the aquatic ecosystem, unless it can be demonstrated that such a discharge will not have an unacceptable adverse impact either individually or in combination with known and/or probable impacts of other activities affecting the ecosystems of concern.

(d) From a national perspective, the degradation or destruction of special aquatic sites, such as filling operations in wetlands, is considered to be among the most severe environmental impacts covered by these Guidelines. The guiding principle should be that degradation or destruction of special sites may represent an irreversible loss of valuable aquatic resources.

* * *

230.10 Restrictions on discharge.

Note: Because other laws may apply to particular discharges and because the Corps of Engineers or State 404 agency may have additional procedural and substantive requirements, a discharge complying with the requirement of these Guidelines will not automatically receive a permit.

Although all requirements in section 230.10 must be met, the compliance evaluation procedures will vary to reflect the seriousness of the potential for adverse impacts on the aquatic ecosystems posed by specific dredged or fill material discharge activities.

(a) Except as provided under section 404(b)(2), no discharge of dredged or fill material shall be permitted if there is a practicable alternative to the proposed discharge which would have less adverse impact on the aquatic ecosystem, so long as the alternative does not have other significant adverse environmental consequences.

(1) For the purpose of this requirement, practicable alternatives include, but are not limited to:

(i)   Activities which do not involve a discharge of dredged or fill material into the waters of the United States or ocean waters;

(ii) Discharges of dredged or fill material at other locations in waters of the United States or ocean waters;

(2) An alternative is practicable if it is available and capable of being done after taking into consideration cost, existing technology, and logistics in light of overall project purposes. If it is otherwise a practicable alternative, an area not presently owned by the applicant which could reasonably be obtained, utilized, expanded or managed in order to fulfill the basic purpose of the proposed activity may be considered.

(3) Where the activity associated with a discharge which is proposed for a special aquatic site (as defined in subpart E) does not require access or proximity to or siting within the special aquatic site in question to fulfill its basic purpose (i.e., is not "water dependent"), practicable alternatives that do not involve special aquatic sites are presumed to be available, unless clearly demonstrated otherwise. In addition, where a discharge is proposed for a special aquatic site, all practicable alternatives to the proposed discharge which do not involve a discharge into a special aquatic site are presumed to have less adverse impact on the aquatic ecosystem, unless clearly demonstrated otherwise.

* * *

(b) No discharge of dredged or fill material shall be permitted if it:

(1) Causes or contributes, after consideration of disposal site dilution and dispersion, to violations of any applicable State water quality standard;

(2) Violates any applicable toxic effluent standard or prohibition under section 307 of the Act;

(3) Jeopardizes the continued existence of species listed as endangered or threatened under the Endangered Species Act of 1973, . . . ;

(4) Violates any requirement imposed by the Secretary of Commerce to protect any marine sanctuary designated under title III of the Marine Protection, Research, and Sanctuaries Act of 1972.

(c) Except as provided under section 404(b)(2), no discharge of dredged or fill material shall be permitted which will cause or contribute to significant degradation of the waters of the United States.

* * *

(d) Except as provided under section 404(b)(2), no discharge of dredged or fill material shall be permitted unless appropriate and practicable steps have been taken which will minimize potential adverse impacts of the discharge on the aquatic ecosystem. Subpart H identifies such possible steps.

* * *

230.75 Actions affecting plant and animal populations.

Minimization of adverse effects on populations of plants and animals can be achieved by:

(a) Avoiding changes in water current and circulation patterns which would interfere with the movement of animals;

(b) Selecting sites or managing discharges to prevent or avoid creating habitat conducive to the development of undesirable predators or species which have a competitive edge ecologically over indigenous plants or animals;

(c) Avoiding sites having unique habitat or other value, including habitat of threatened or endangered species;

(d) Using planning and construction practices to institute habitat development and restoration to produce a new or modified environmental state
of higher ecological value by displacement of some or all of the existing
environmental characteristics. Habitat development and restoration
techniques can be used to minimize adverse impacts and to compensate
for destroyed habitat. Use techniques that have been demonstrated to
be effective in circumstances similar to those under consideration wher-
ever possible. Where proposed development and restoration techniques have not yet advanced to the pilot demonstration stage, initiate
their use on a small scale to allow corrective action if unanticipated adverse impacts occur;

(e) Timing discharge to avoid spawning or migration seasons and other bi-
ologically critical time periods;

(f) Avoiding the destruction of remnant natural sites within areas already
affected by development.

Corps Regulations 33 CFR Parts 320–332 (excerpts)

§ 320.4 General policies for evaluating permit applications.

(a) Public Interest Review.

(1) The decision whether to issue a permit will be based on an evaluation of the probable impacts, including cumulative impacts, of
the proposed activity and its intended use on the public interest.
Evaluation of the probable impact which the proposed activity
may have on the public interest requires a careful weighing of all
those factors which become relevant in each particular case. The
benefits which reasonably may be expected to accrue from the proposal must be balanced against its reasonably foreseeable detri-
ments. The decision whether to authorize a proposal, and if so, the
conditions under which it will be allowed to occur, are therefore
determined by the outcome of this general balancing process. That
decision should reflect the national concern for both protection
and utilization of important resources. All factors which may be
relevant to the proposal must be considered including the cumulative effects thereof: among those are conservation, economics, aesthetics, general environmental concerns, wetlands, historic properties, fish and wildlife values, flood hazards, floodplain values, land
use, navigation, shore erosion and accretion, recreation, water sup-

ply and conservation, water quality, energy needs, safety, food and fiber production, mineral needs, considerations of property ownership and, in general, the needs and welfare of the people. For activities involving 404 discharges, a permit will be denied if the discharge that would be authorized by such permit would not comply with the Environmental Protection Agency's 404(b)(1) guidelines. Subject to the preceding sentence and any other applicable guidelines and criteria (see §§320.2 and 320.3), a permit will be granted unless the district engineer determines that it would be contrary to the public interest.

* * *

§ 328.3 Definitions.

For the purpose of this regulation these terms are defined as follows:

(a) The term waters of the United States means

   (1) All waters which are currently used, or were used in the past, or may be susceptible to use in interstate or foreign commerce, including all waters which are subject to the ebb and flow of the tide;

   (2) All interstate waters including interstate wetlands;

   (3) All other waters such as intrastate lakes, rivers, streams (including intermittent streams), mudflats, sandflats, wetlands, sloughs, prairie potholes, wet meadows, playa lakes, or natural ponds, the use, degradation or destruction of which could affect interstate or foreign commerce including any such waters:

      (i) Which are or could be used by interstate or foreign travelers for recreational or other purposes; or

      (ii) From which fish or shellfish are or could be taken and sold in interstate or foreign commerce; or

      (iii) Which are used or could be used for industrial purpose by industries in interstate commerce;

   (4) All impoundments of waters otherwise defined as waters of the United States under the definition;

   (5) Tributaries of waters identified in paragraphs (a) (1) through (4) of this section;

(6) The territorial seas;

(7) Wetlands adjacent to waters (other than waters that are themselves wetlands) identified in paragraphs (a) (1) through (6) of this section.

(8) Waters of the United States do not include prior converted cropland. Notwithstanding the determination of an area's status as prior converted cropland by any other Federal agency, for the purposes of the Clean Water Act, the final authority regarding Clean Water Act jurisdiction remains with EPA.

Waste treatment systems, including treatment ponds or lagoons designed to meet the requirements of CWA (other than cooling ponds as defined in 40 CFR 423.11(m) which also meet the criteria of this definition) are not waters of the United States.

(b) The term wetlands means those areas that are inundated or saturated by surface or ground water at a frequency and duration sufficient to support, and that under normal circumstances do support, a prevalence of vegetation typically adapted for life in saturated soil conditions. Wetlands generally include swamps, marshes, bogs, and similar areas.

(c) The term adjacent means bordering, contiguous, or neighboring. Wetlands separated from other waters of the United States by man-made dikes or barriers, natural river berms, beach dunes and the like are "adjacent wetlands."

* * *

§ 332.1 Purpose and general considerations.

(a) Purpose.

(1) The purpose of this part is to establish standards and criteria for the use of all types of compensatory mitigation, including on-site and off-site permittee-responsible mitigation, mitigation banks, and in-lieu fee mitigation to offset unavoidable impacts to waters of the United States authorized through the issuance of Department of the Army (DA) permits pursuant to section 404 of the Clean Water Act (33 U.S.C. 1344) and/or sections 9 or 10 of the Rivers and Harbors Act of 1899 (33 U.S.C. 401, 403).

* * *

(c) Sequencing.

   (1) Nothing in this section affects the requirement that all DA permits subject to section 404 of the Clean Water Act comply with applicable provisions of the Section 404(b)(1) Guidelines at 40 CFR part 230.

   (2) Pursuant to these requirements, the district engineer will issue an individual section 404 permit only upon a determination that the proposed discharge complies with applicable provisions of 40 CFR part 230, including those which require the permit applicant to take all appropriate and practicable steps to avoid and minimize adverse impacts to waters of the United States. Practicable means available and capable of being done after taking into consideration cost, existing technology, and logistics in light of overall project purposes. Compensatory mitigation for unavoidable impacts may be required to ensure that an activity requiring a section 404 permit complies with the Section 404(b)(1) Guidelines.

   (3) Compensatory mitigation for unavoidable impacts may be required to ensure that an activity requiring a section 404 permit complies with the Section 404(b)(1) Guidelines. During the 404(b)(1) Guidelines compliance analysis, the district engineer may determine that a DA permit for the proposed activity cannot be issued because of the lack of appropriate and practicable compensatory mitigation options.

                              * * *

(f) Relationship to other guidance documents.

   (1) This part applies instead of the "Federal Guidance for the Establishment, Use, and Operation of Mitigation Banks," which was issued on November 28, 1995, the "Federal Guidance on the Use of In-Lieu Fee Arrangements for Compensatory Mitigation Under Section 404 of the Clean Water Act and Section 10 of the Rivers and Harbors Act," which was issued on November 7, 2000, and Regulatory Guidance Letter 02–02, "Guidance on Compensatory Mitigation Projects for Aquatic Resource Impacts Under the Corps Regulatory Program Pursuant to Section 404 of the Clean Water Act and Section 10 of the Rivers and Harbors Act of 1899" which was issued on December 24, 2002. These guidance documents are no longer to be used as compensatory mitigation policy in the Corps Regulatory Program.

(2) In addition, this part also applies instead of the provisions relating to the amount, type, and location of compensatory mitigation projects, including the use of preservation, in the February 6, 1990, Memorandum of Agreement (MOA) between the Department of the Army and the Environmental Protection Agency on the Determination of Mitigation Under the Clean Water Act Section 404(b)(1) Guidelines. All other provisions of this MOA remain in effect.

§ 332.2 Definitions.

For the purposes of this part, the following terms are defined:

* * *

Compensatory mitigation means the restoration (re-establishment or rehabilitation), establishment (creation), enhancement, and/or in certain circumstances preservation of aquatic resources for the purposes of offsetting unavoidable adverse impacts which remain after all appropriate and practicable avoidance and minimization has been achieved.

Credit means a unit of measure (e.g., a functional or areal measure or other suitable metric) representing the accrual or attainment of aquatic functions at a compensatory mitigation site. The measure of aquatic functions is based on the resources restored, established, enhanced, or preserved.

In-lieu fee program means a program involving the restoration, establishment, enhancement, and/or preservation of aquatic resources through funds paid to a governmental or non-profit natural resources management entity to satisfy compensatory mitigation requirements for DA permits. Similar to a mitigation bank, an in-lieu fee program sells compensatory mitigation credits to permittees whose obligation to provide compensatory mitigation is then transferred to the in-lieu program sponsor. However, the rules governing the operation and use of in-lieu fee programs are somewhat different from the rules governing operation and use of mitigation banks. The operation and use of an in-lieu fee program are governed by an in-lieu fee program instrument.

Mitigation bank means a site, or suite of sites, where resources (e.g., wetlands, streams, riparian areas) are restored, established, enhanced, and/or preserved for the purpose of providing compensatory mitigation for impacts authorized by DA permits. In general, a mitigation bank sells compensatory mitigation credits to permittees whose obligation to provide compensatory mitigation is then transferred to the mitigation bank sponsor. The operation and use of a mitigation bank are governed by a mitigation banking instrument.

Performance standards are observable or measurable physical (including hydrological), chemical and/or biological attributes that are used to determine if a compensatory mitigation project meets its objectives.

Permittee-responsible mitigation means an aquatic resource restoration, establishment, enhancement, and/or preservation activity undertaken by the permittee (or an authorized agent or contractor) to provide compensatory mitigation for which the permittee retains full responsibility.

Service area means the geographic area within which impacts can be mitigated at a specific mitigation bank or an in-lieu fee program, as designated in its instrument.

Watershed means a land area that drains to a common waterway, such as a stream, lake, estuary, wetland, or ultimately the ocean.

Watershed approach means an analytical process for making compensatory mitigation decisions that support the sustainability or improvement of aquatic resources in a watershed. It involves consideration of watershed needs, and how locations and types of compensatory mitigation projects address those needs. A landscape perspective is used to identify the types and locations of compensatory mitigation projects that will benefit the watershed and offset losses of aquatic resource functions and services caused by activities authorized by DA permits. The watershed approach may involve consideration of landscape scale, historic and potential aquatic resource conditions, past and projected aquatic resource impacts in the watershed, and terrestrial connections between aquatic resources when determining compensatory mitigation requirements for DA permits.

Watershed plan means a plan developed by federal, tribal, state, and/or local government agencies or appropriate non-governmental organizations, in consultation with relevant stakeholders, for the specific goal of aquatic resource restoration, establishment, enhancement, and preservation. A watershed plan addresses aquatic resource conditions in the watershed, multiple stakeholder interests, and land uses. Watershed plans may also identify priority sites for aquatic resource restoration and protection. Examples of watershed plans include special area management plans, advance identification programs, and wetland management plans.

Clean Water Act Guidance Documents (excerpts)

Clean Water Act Jurisdiction Following the U.S. Supreme Court's Decision in *Rapanos v. United States* and *Carabell v. United States*

This memorandum provides guidance to EPA regions and U.S. Army Corps of Engineers ["Corps"] districts implementing the Supreme Court's decision in the consolidated cases *Rapanos v. United States* and *Carabell v. United States* (herein referred to simply as "Rapanos") which address the jurisdiction over waters of the United States under the Clean Water Act. The chart below summarizes the key points contained in this memorandum. This reference tool is not a substitute for the more complete discussion of issues and guidance furnished throughout the memorandum.

Summary of Key Points

The agencies will assert jurisdiction over the following waters:

- Traditional navigable waters
- Wetlands adjacent to traditional navigable waters
- Nonnavigable tributaries of traditional navigable waters that are relatively permanent where the tributaries typically flow year-round or have continuous flow at least seasonally (e.g., typically three months)
- Wetlands that directly abut such tributaries

The agencies will decide jurisdiction over the following waters based on a fact-specific analysis to determine whether they have a significant nexus with a traditional navigable water:

- Nonnavigable tributaries that are not relatively permanent
- Wetlands adjacent to nonnavigable tributaries that are not relatively permanent
- Wetlands adjacent to but that do not directly abut a relatively permanent nonnavigable tributary

Summary of Key Points (*continued*)

The agencies generally will not assert jurisdiction over the following features:

- Swales or erosional features (e.g., gullies, small washes characterized by low volume, infrequent, or short duration flow)
- Ditches (including roadside ditches) excavated wholly in and draining only uplands and that do not carry a relatively permanent flow of water

The agencies will apply the significant nexus standard as follows:

- A significant nexus analysis will assess the flow characteristics and functions of the tributary itself and the functions performed by all wetlands adjacent to the tributary to determine if they significantly affect the chemical, physical, and biological integrity of downstream traditional navigable waters.
- Significant nexus includes consideration of hydrologic and ecologic factors.

* * *

Regulatory Guidance Letter 95-01

SUBJECT: Guidance on Individual Permit Flexibility for Small Landowners

* * *

Memorandum for the Field

In order to clearly affirm the flexibility afforded to small landowners under Section 404 of the Clean Water Act, this policy clarifies that for discharges of dredged or fill material affecting up to two acres of non-tidal wetlands for the construction or expansion of a home or farm building, or expansion of a small business, it is presumed that alternatives located on property not currently owned by the applicant are not practicable under the Section 404(b)(1) Guidelines.

Specifically, for those activities involving discharges of dredged or fill material affecting up to two acres into jurisdictional wetlands for:

1. the construction or expansion of a single family home and attendant features, such as a driveway, garage, storage shed, or septic field;

2. the construction or expansion of a barn or other farm building; or

3. the expansion of a small business facility;

which are not otherwise covered by a general permit, it is presumed that alternatives located on property not currently owned by the applicant are not practicable under the Section 404(b)(1) Guidelines. The Guidelines' requirements to appropriately and practicably minimize and compensate for any adverse environmental impacts of such activities remain.

Chapter 2

1. One might argue that *The Pelican Brief*, based on the book by John Grisham and starring Denzel Washington and Julia Roberts, is an exception. In the movie, two U.S. Supreme Court justices are assassinated as part of a conspiracy to allow an evil oil company to despoil Louisiana wetlands. Julia Roberts's character (Darby) is a law student who writes a paper (the "Pelican Brief") for her law professor that outlines the company's plot. But the movie never provides any legal details (was it the Clean Water Act or the Endangered Species Act at issue?) and its title is misleading. Darby merely wrote a memo to her professor, not a brief, which is a written submission that a lawyer provides a court. The movie should have been called *The Pelican Memorandum*, but this likely would have reduced its receipts.

2. The common law generally refers to law developed through judicial decisions and precedents, in contrast to law derived from legislative acts or regulations issued by executive branch agencies. For a more nuanced view of the common law, see Fletcher and Sheppard (2005).

3. Although most agencies are within the executive branch, there are some legislative agencies, such as the Government Accountability Office and the Congressional Budget Office, and judicial agencies, such as the U.S. Sentencing Commission.

4. Amy Skilbred, the plaintiff who had visited Sri Lanka, stated she had no current plans to return, in part because of an ongoing civil war. The Sri Lankan government's battle with the Tamil Tigers appears to have finally ended in 2009, seventeen years after the Court's decision in *Lujan*.

5. For example, an agency's interpretation that first appears in an appellate brief (known as "post hoc rationalization of counsel") will not be afforded *Chevron* deference.

Chapter 3

1. Technically, some wetlands can be considered deserts. For example, tundra in Alaska that receives less than 10 inches of rainfall annually is a desert under some definitions. In warmer months, these areas may pond and exhibit wetland conditions as the underlying permafrost maintains water at or near the surface (Hall et al., 1994).

2. In the 1978 case of *Regents of the University of California v. Bakke*, the Supreme Court grappled with the issue of race in university admissions programs and split 4-4-1. Justice Powell was the dispositive vote, concluding that race could be one of several factors. Although no other justice joined his concurrence, his approach was used by universities and eventually was endorsed by a majority of the Supreme Court in 2003.

3. To strain the analogy further: just as in Tony Soprano's world, the wetland world has bad guys (Mr. Rapanos: "Destroy the records or I'll destroy you"), and federal law enforcement lurks in the background. Characters also get whacked over time: witness the demise of the Migratory Bird Rule and the resulting evisceration of protections for isolated wetlands such as vernal pools.

4. For example, the plurality suggested that "seasonal" rivers could be considered relatively permanent. *Rapanos*, 547 U.S. at 732 n.5. What does "seasonal" mean? The agencies gave it a plain meaning interpretation: a season is technically three months in length and thus a seasonal flow is typically three months.

5. References to "muddy waters" seem to be most common. For example, see Lawrence R. Liebesman and Rafe Petersen, "Corps and EPA Guidance Attempts to Clarify Clean Water Act Jurisdiction 'Muddied' by the Supreme Court in *Rapanos v. United States*," http://www.martindale.com/environmental-law/article _Holland-Knight-LLP_302524.htm.

Chapter 4

1. To be specific: Clean Water Act section 301 prohibits all point source discharges of pollutants into navigable waters without a permit. Many states have different permit requirements, which can be broader. In Florida, for example, an environmental resource permit is required for *any* land-use or construction activity that could affect wetlands, not just point source discharges.

2. A state can object to the issuance of a NWP within its jurisdiction. In such cases, the NWP is not available to permit applicants, and they must then seek an individual permit.

Chapter 5

1. In *Zabel v. Tabb*, a developer sought to fill 11 acres of Boca Ciega Bay in Florida for a trailer park. For the first time in its history, the Corps denied a Rivers and Harbor Act permit based on environmental grounds (as opposed to navigational reasons). The Fifth Circuit Court of Appeals upheld the Corps' authority, stating that

[w]e hold that nothing in the statutory structure compels the Secretary [of the Army] to close his eyes to all that others see or think they see. The establishment was entitled, if not required, to consider ecological factors and, being persuaded by them, to deny that which might have been granted routinely five, ten, or fifteen years ago before man's explosive increase made all, including

Congress, aware of civilization's potential destruction from breathing its own polluted air and drinking its own infected water and the immeasurable loss from a silent-spring-like disturbance of nature's economy.

2. Under revised elevation MOAs in 1992, the EPA, the FWS, and the National Oceanic and Atmospheric Administration can elevate policy issues and individual permit decisions. However, a policy elevation will not delay decisions regarding individual permits, and elevations of individual permit decisions are limited to cases in which an aquatic resource of national importance (known as an ARNI) is involved.

3. Before a grand jury, President Clinton stated, "It depends on what the meaning of the word 'is' is. If the—if he—if 'is' means is and never has been, that is not—that is one thing. If it means there is none, that was a completely true statement. . . . Now, if someone had asked me on that day, are you having any kind of sexual relations with Ms. Lewinsky, that is, asked me a question in the present tense, I would have said no. And it would have been completely true." Clinton was impeached by the U.S. House of Representatives for perjury to the grand jury (as well as for obstruction of justice) and had his Arkansas law license suspended.

4. I would suggest not (Gardner, 1990). Indeed, the MOA became well respected over time, so much so that changes to it were met with howls of objections from some environmental groups (Gardner, 2002).

5. It also helped that Senator Lott retired from the Senate in December 2007.

Chapter 6

1. In Spanish, the word *nutria* refers to the otter, which has a much better reputation.

Chapter 7

1. Indeed, the first wetland mitigation bank was the Fina LaTerre Mitigation Bank in Louisiana, which was initially opened as a single-user bank for the Tenneco Oil Company in 1985 (ELI and IWR, 1994; Roberston, 2008).

2. Initially, 15 percent of credits were released after a conservation easement was placed on the site. Removal of the invasive melaleuca resulted in a release of 25 percent of the credits, and another 40 percent were awarded when the site was graded. Successful planting yielded another 10 percent of the credits, and the final 10 percent was contingent on successful monitoring (Reiss et al., 2007).

3. At the conclusion of one lengthy interagency review process, a California banker's closing salutation on the letter transmitting the final agreement was not "Sincerely" but "Hallelujah" (Gardner and Radwan, 2005).

4. To close the loop yet again, in 2003 the EPA, the Corps, and the Federal Highway Administration issued guidance on how to fulfill TEA-21's preference for mitigation bank credits.

5. Accordingly, when *Rapanos* reached the U.S. Supreme Court, the National Mitigation Banking Association filed an amicus brief, aligning itself with environmental groups on the issue of the constitutionality of federal jurisdiction over waters that are not traditionally navigable.

6. Really. Location of a mitigation bank is important in terms of financial viability (it should be near an area of development), ecological contributions (these will depend on its placement in the watershed), and long-term maintenance (this will be affected by current and foreseeable adjacent land uses).

Chapter 8

1. The mitigation properties included conservation bank lands, rather than wetland mitigation bank lands. Conservation banks are similar to wetland banks, but focus on endangered species habitat.

Chapter 9

1. For example, the proposed rule stated that mitigation plans for permittee-responsible mitigation and mitigation banks should contain a description of the mitigation objectives (type and amount of compensatory mitigation to be provided); site selection process (including watershed needs); legal instruments such as conservation easements to protect the site for the long term; baseline information of the mitigation and impact sites; credit calculations; work plans (construction methods, plants to be removed and planted); ecologically based performance standards; maintenance and monitoring plans; and long-term management and adaptive management procedures.

2. After conversations with agency officials involved in the OMB discussions, I floated the idea in public. I wrote an editorial, suggesting that the agencies keep in-lieu fee mitigation as an option, but delink it from third-party mitigation, treating it as a form of permittee-responsible mitigation. The piece was entitled "Reconsidering In-Lieu Fees." A Chicago-area mitigation banker responded with a pithy e-mail entitled "reconsidering your sanity."

3. One ivory tower resident suggested such an idealistic approach in 1996. See Royal C. Gardner, "Banking on Entrepreneurs: Wetlands, Mitigation Banking, and Takings," 81 *Iowa Law Review* 527 (1996).

Chapter 10

1. A court has a great deal of discretion in fashioning injunctive remedies, including the requirement of modern-day scarlet letters. The most vivid illustration comes from *United States v. Van Leuzen*, where the Court ordered the violator to restore the site and erect a 10-foot by 20-foot billboard to explain to passing cars and pedestrians that he had violated the Clean Water Act and was removing the fill and restoring the site.

2. See, for example, Gardner (2000).

3. Because of permit reporting requirements, such a case is relatively simple. Section 402 permittees typically have to provide discharge monitoring reports (DMRs) to the EPA and/or state agency. A prospective plaintiff can compare the level of pollutants in the DMR with the acceptable level in the permit. If the former exceeds the latter, the case is a good candidate for summary judgment, in which the court can rule without holding a full trial.

Chapter 11

1. The state of South Carolina used its power of eminent domain to take private property to establish the SCDOT Black River Mitigation Bank as a single-user bank to offset impacts associated with state transportation projects. The landowner, Faulkenberry, objected to the compensation offered, so the matter went to a jury trial. Faulkenberry argued that he was entitled to compensation based on what the land would be worth as a wetland mitigation bank. The state countered that such a valuation would be inappropriate because it was unlikely that Faulkenberry would have been able to set up a bank to sell credits to private developers. The jury agreed with Faulkenberry and awarded him approximately $2.4 million.

2. The EPA was acting under Clean Water Act section 402, rather than section 404, since the oil brine was considered waste. The court's decision applies with equal force to section 404 actions, at least for purposes of the geographic jurisdiction of the Clean Water Act and general takings principles.

3. Courts have recognized a futility exception to the general rule. A takings plaintiff does not necessarily need to apply for a permit multiple times: "once it becomes clear that the agency lacks the discretion to permit any development . . . a takings claim is likely to have ripened." *Palazzolo v. Rhode Island*, 533 U.S. 606, 620 (2001).

4. The regulated activity was the movement of soil that resulted in the discharge of dredged or fill material; the addition of caskets would not constitute a discharge of dredged or fill material.

5. Some enterprising property owners have attempted to claim as a tax deduction the loss of value in a property as a result of wetland regulation. When the deduction is disallowed, the disappointed taxpayer has three options: refuse to pay and litigate in U.S. Tax Court; pay and institute a refund action in U.S. District Court; or pay and institute a refund action in the U.S. Court of Federal Claims. Property owners who have attempted to take such deductions have been uniformly unsuccessful in subsequent litigation.

6. That almost tells you all you need to know: A Philadelphia attorney familiar with wetland regulations purchases property in Florida. He is going to lose.

7. In its opinion, the Court first disposed of a procedural issue. In 1990, during the pendency of the case before the South Carolina Supreme Court, South Carolina amended the act to provide for special permits for occupiable structures. The

Coastal Council argued to the U.S. Supreme Court that Lucas must now apply for the special permit, and a decision be made on that application, before his takings claim could be considered appropriate for judicial review. The Court concluded that Lucas had presented a ripe claim for a temporary taking with respect to the time period between 1988 and 1990.

8. Judge Loren Smith was viewed as very friendly toward private property claims and found that a taking had occurred in wetland cases such as *Florida Rock*, *Loveladies Harbor v. United States*, *Bowles v. United States*, and *Cooley v. United States* (which was later reversed).

Chapter 12

1. It is speculated that the proposal was killed at the request of oil and gas lobbyists who (exercising their First Amendment rights) met with Vice President Cheney as part of the Bush administration's Energy Task Force (Kennedy, 2010). MMS also had a reputation for lax enforcement and ethical lapses. It failed to enforce a regulation requiring BP to submit testing data regarding the effectiveness of blowout preventers (Barstow et al., 2010). In addition, the agency reportedly allowed oil companies to explore and drill in the Gulf of Mexico in contravention of the Endangered Species Act and the Marine Mammal Protection Act (Urbina, 2010). An Interior Inspector General's report released in May 2010 found that MMS employees routinely accepted gifts from the oil and gas industry, including a flight on a private jet to Atlanta to watch Louisiana State University's football team play in the Peach Bowl (Office of Inspector General, 2010). LSU won the game, but the public ultimately lost.

2. This recommendation does not mean to suggest that states should be excluded from wetland regulation. They can administer a joint program or even assume responsibility for the federal program as Michigan and New Jersey have done.

3. Some combinations, however, are currently prohibited. The Corps and the FWS do not allow stacking of wetland and endangered species credits on the same parcel. And after a recent controversy in North Carolina involving wetlands and water quality credits, there is a stacking moratorium in that state (Program Evaluation Division, NC General Assembly, 2009). On the other hand, the stacking of carbon sequestration credits with other types of credits may have more support.

4. The call for transparency from the environmental groups also invoked the December 2009 "Open Government Directive" issued by the Office of Management and Budget. This OMB memo encourages agencies to disseminate useful information through the Web without waiting for specific Freedom of Information Act requests. Note, however, that the OMB memo is not a regulation. It is guidance and thus does not legally bind agencies.

SELECTED REFERENCES
AND FURTHER READING

Chapter 1

Comprehensive Everglades Restoration Plan (CERP). (2010). *The Journey to Restore America's Everglades*, www.evergladesplan.org/index.aspx.

Douglas, M.S. (1988). *The Everglades: River of Grass* (rev. ed.). Sarasota, FL: Pineapple Press.

Doyle, A.C. (1959). *Hound of the Baskervilles*. New York, NY: Random House Children's Books.

Ducks Unlimited. (2010). www.ducks.org.

Grunwald, M. (2006). *The Swamp: The Everglades, Florida, and the Politics of Paradise*. New York: Simon and Schuster,

Kenny, A. (2006). "Ecosystem Services in the New York City Watershed." Ecosystem Marketplace, www.ecosystemmarketplace.com/pages/dynamic/article.page.php?page_id=4130&section=home.

Kolva, J.R. (1996). "Effects of the Great Midwest Flood of 1993 on Wetlands." National Water Summary on Wetland Resources, U. S. Geological Survey Water Supply Paper 2425.

Labyrinth. (1999). DVD. Dir. Jim Henson. Perf. David Bowie and Jennifer Connelly. 1989. Sony Pictures.

Leovy v. United States (1900) 177 U.S. 621 U.S. Supreme Court.

Littleton, S.C. (2005). *Gods, Goddesses, and Mythology*. Tarrytown, NY: Marshall Cavendish Corp.

Mitsch, W.J. and J.G. Gosselink. (2007). *Wetlands*. Hoboken, NJ: John Wiley & Sons, Inc.

Monaghan, P. (2004). *The Encyclopedia of Celtic Mythology and Folklore*. New York: Facts on File.

Monty Python and the Holy Grail (Special Edition). (2001). DVD. Dir. Terry Jones. Perf. John Cleese and Graham Chapman. 1975. Sony Pictures.

The Princess Bride. DVD. (2007). Dir. Rob Reiner. Perf. Cary Elwes, Mandy Patinkin, and Robin Wright Penn. 1987. MGM.

Ramsar Convention Secretariat. (2010a). "Shoreline stabilisation & storm protection." www.ramsar.org/pdf/info/services_03_e.pdf.

Ramsar Convention Secretariat. (2010b). "World Wetlands Day." www.ramsar.org/wwd/wwd_index.htm.

Swamp Thing. (1972). New York: DC Comic Group.

Tiner, R.W., J.Q. Swords, and B.J. McClain. (2002). *Wetland Status and Trends for the Hackensack Meadowlands. An Assessment Report from the U.S. Fish and Wildlife Service's National Wetlands Inventory Program*. Hadley, MA: U.S. Fish and Wildlife Service, Northeast Region.

U.S. Environmental Protection Agency. (2010a). "America's Wetland Month." www.epa.gov/owow/wetlands/awm/.

U.S. Environmental Protection Agency. (2010b). *Watershed Academy Web*. http://cfpub.epa.gov/watertrain/index.cfm.

U.S. Fish and Wildlife Service. (2009). *Birding in the United States: A Demographic and Economic Analysis, Addendum to the 2006 National Survey of Fishing, Hunting, and Wildlife-Associated Recreation*. Arlington, VA.

Vileisis, A. (1997). *Discovery of the Unknown Landscape: A History of America's Wetlands*. Washington, D.C: Island Press.

"Wetlands and Marshes." (2010). *The Thain's Book: An Encyclopedia of Middle Earth and Numenor*. www.tuckborough.net/marshes.html.

When the Levees Broke: A Requiem in Four Acts. (2006). DVD. Dir. Spike Lee. HBO Home Video.

The Winged Scourge. (2005). DVD. *Walt Disney Treasures—On the Front Lines*. Buena Vista Home Entertainment.

Zion, G. (1957). *Dear Garbage Man*. New York: HarperCollins Publishers.

Chapter 2

Adler, R., J. Landman, and D. Cameron. (1993). *The Clean Water Act 20 Years Later*. Washington, D.C: Island Press.

Administrative Procedure Act. (2006). U.S. Code 5:553.

Barringer, F. (2003). "U.S. Won't Narrow Wetlands Protection," *New York Times*, Dec. 17.

Chevron U.S.A., Inc. v. Natural Resources Defense Council (1984) 467 U.S. 837, U.S. Supreme Court.

Coquillette, D. (1979). "Mosses from an Old Manse: Another Look at Some Historic Property Cases About the Environment." *Cornell Law Review* 64, 761–820.

Editorial. (2001). "Two BU.S.hes and the Everglades." *New York Times*, Nov. 23.

Executive Order 12866, as amended by E.O. 13258 and E.O. 13422 on Regulatory Planning and Review (January 18, 2007).

Exxon Shipping Co. v. Baker. (2008). 128 S. Ct. 2605, U.S. Supreme Court.

Fletcher, G.P. and S. Sheppard. (2005). *American Law in a Global Context*. New York: Oxford University Press.

Fuhr, J. (2010). "*Connecticut v. AEP*: The New Normal?" *Natural Resources & Environment* 24:4, 58–59.

Gardner, R. (1990). "Public Participation and Wetland Regulation," *UCLA Journal of Environmental Law and Policy* 10, 1–39.

Gwaltney v. Chesapeake Bay Foundation. (1987). 484 U.S. 49, U.S. Supreme Court.

In re Cheney. (2005). 406 F.3d 723, D.C. Circuit Court of Appeals.

Kagan, E. (2001). "Presidential Administration." *Harvard Law Review* 114, 2245–2385.

Kalen, S. (2008). "Changing Administrations and Environmental Guidance Documents." *Natural Resources & Environment* 23:3, 13–17.

Lujan v. Defenders of Wildlife (1992) 504 U.S. 555, U.S. Supreme Court.

Municipality of Anchorage v. United States (1992) 980 F.2d 1320, Ninth Circuit Court of Appeals.

Percival, R. (2001). "Presidential Management of the Administrative State: The Not-So-Unitary Executive." *Duke Law Journal* 51, 963–1013.

Pierce, R. (1995). "Seven Ways to Deossify Agency Rulemaking." *Administrative Law Review* 47, 59–95.

Pierce, R., S. Shapiro, and P. Verkuil. (2009). *Administrative Law and Process*. New York: Thomson/West.

United States v. Mead Corp. (2001). 533 U.S. 218, U.S. Supreme Court.

U.S. Army Corps of Engineers Waterways Experiment Station. (1987). *Corps of Engineers Wetlands Delineation Manual*, Technical Report Y-87-1. Vicksburg, MS.

U.S. Government Accountability Office. (2005a). "Waters and Wetlands: Corps of Engineers Needs to Better Support Its Decisions for Not Asserting Jurisdiction." GAO-05-870.

U.S. Government Accountability Office. (2005b). "Wetlands Protection: Corps of Engineers Does Not Have an Effective Oversight Approach to Ensure That Compensatory Mitigation Is Occurring." GAO-05-898.

U.S. Government Accountability Office. (2008). *Principles of Federal Appropriations Law*, 3rd ed., vol III. GAO-08-978SP.

Worth, R. (1999). "Asleep on the Beat." *Washington Monthly* 31,11.

Chapter 3

Cowardin, L.M., V. Carter, F.C. Golet, and E.T. LaRoe. (1979). *Classification of Wetlands and Deepwater Habitats of the United States*. Washington, D.C.: U.S. Fish and Wildlife Service.

Craig, R. (2009). *The Clean Water Act and the Constitution*. Washington, D.C.: Island Press.

Environmental Defense Fund and World Wildlife Fund. (1992). *How Wet Is a Wetland? The Impacts of the Proposed Revisions to the Federal Wetlands Delineation Manual*. Laurel, MD.

Feingold, R. (2007). "Restoring Federal Jurisdiction Over 'Waters of the United States'." *National Wetlands Newsletter* 29:3, 3–4.

Goldman-Carter, J. (2005). "Isolated Wetland Legislation: Running the Rapids at the State Capitol." *National Wetlands Newsletter* 27:3, 27-29.

Gonzales v. Raich. (2006). 545 U.S. 1 U.S. Supreme Court.

Hall, J.V., W.E. Frayer, and B.O. Wilen. (1994). *Status of Alaska Wetlands.* Anchorage, AK: U.S. Fish and Wildlife Service Alaska Region.

KU.S.ler, J. (2004). "The SWANCC Decision: State Regulation of Wetlands to Fill the Gap." Association of State Wetland Managers, Berne, NY. www.aswm.org /fwp/swancc/aswm-int.pdf.

Lewis, W. (2001). *Wetlands Explained.* Oxford University Press, New York, NY.

National Research Council. (1995). *Wetlands: Characteristics and Boundaries.* Washington, D.C.: National Academies Press.

NRDC v. Callaway. (1975). 392 F.Supp. 685 U.S. District Court for the District of Columbia.

Rapanos v. U.S.. (2006). 547 U.S. 715 U.S. Supreme Court.

Robertson, M.M. (2004). "Drawing Lines in Water: Entrepreneurial Wetland Mitigation Banking and the Search for Ecosystem Service Markets." Ph.D. thesis, University of Wisconsin.

Solid Waste Agency of Northern Cook County v. U.S. Army Corps of Engineers (2001) 531 U.S. 159 U.S. Supreme Court.

U.S. Army Corps of Engineers Waterways Experiment Station. (1987). *Corps of Engineers Wetlands Delineation Manual,* Technical Report Y-87-1. Vicksburg, MS.

United States v. Riverside Bayview Homes. (1985). 474 U.S. 121 U.S. Supreme Court.

U.S. Environmental Protection Agency and U.S. Department of the Army. (2008). "Clean Water Act Jurisdiction Following the U.S. Supreme Court's Decision in *Rapanos v. United States* & *Carabell v. United States.*" Washington, D.C..

Wickard v. Filburn. (1942). 317 U.S. 111 U.S. Supreme Court.

Wroth, K., ed. (2007). *The Supreme Court and the Clean Water Act: Five Essays.* Royalton, VT: Vermont Law School's Land Use Institute,

Chapter 4

Avoyelles Sportsmen's League, Inc. v. Marsh. (1983). 715 F.2d 897 U.S. Court of Appeals for the Fifth Circuit.

Bishop, T.S. et al. (2004). "Counting the Hands on *Borden Ranch.*" *Environmental Law Reporter* 34, 10040–44.

Borden Ranch Partnership v. U.S. Army Corps of Engineers. (2001). 261 F.3d 810 U.S. Court of Appeals for the Ninth Circuit.

Coeur Alaska, Inc. v. Southeast Alaska Conservation Council. (2009). 129 S.Ct. 2458 U.S. Supreme Court.

Copland, C. (2009). "Controversies over Redefining 'Fill Material' Under the Clean Water Act." Washington, D.C.: Congressional Research Service.

FU.S.chino, J. (2007). "Mountaintop Coal Mining and the Clean Water Act: The Fight over Nationwide Permit 21." *Boston College Environmental Affairs Law Review* 34, 179–206.

Gardner, R.C. (1998). "Casting Aside the Tulloch Rule." *National Wetlands Newsletter* 20:5, 5–6, 24.

National Mining Association v. U.S. Army Corps of Engineers. (1998). 145 F.3d 1399 U.S. Court of Appeals for the D.C. Circuit.

Ohio Valley Environmental Coalition v. Aracoma Coal Co. (2009). 556 F.3d 177 U.S. Court of Appeals for the Fourth Circuit.

Richardson, C.J. (1983). "Pocosins: Vanishing Wastelands or Valuable Wetlands?" *BioScience* 33:10, 626–633.

Taylor, W.E. and K.L. Geoffrey. (2005). "General and Nationwide Permits," in K.D. Connolly et al., eds., *Wetlands Law and Policy: Understanding Section 404*, pp151–190. Chicago: American Bar Association.

U.S. Army Corps of Engineers. (1984). *Agricultural Conversion: Fifth Circuit Decision in Avoyelles vs Marsh*. RGL 84-05.

U.S. Army Corps of Engineers. (2010). "Nationwide Permits Information." www.usace.army.mil/cecw/pages/nw_permits.aspx.

U.S. Environmental Protection Agency. (2005). *Mountaintop Mining/Valley Fills in Appalachia, Final Programmatic Impact Statement*. Philadelphia, PA.

U.S. Environmental Protection Agency. (2010a). "Further Revisions to the Clean Water Act Regulatory Definition of Discharge of Dredged Material." http://water.epa.gov/lawsregs/lawsguidance/cwa/dredging/2001rule.cfm.

U.S. Environmental Protection Agency. (2010b). "Mid-Atlantic Mountaintop Mining." www.epa.gov/region3/mtntop/.

Chapter 5

Bersani v. U.S. Environmental Protection Agency. (1988). 850 F.2d 36 U.S. Court of Appeals for the Second Circuit.

Blumm, M.C., and D.B. Zaleha. (1989). "Federal Wetlands Protection under the Clean Water Act: Regulatory Ambivalence, Intergovernmental Tension, and a Call for Reform." *University of Colorado Law Review* 60, 695–772.

Fund for Animals v. Rice. (1996). 85 F.3d 535 U.S. Court of Appeals for the Eleventh Circuit.

Gardner, R.C. (1990). "The Army-EPA Mitigation Agreement: No Retreat from Wetlands Protection." *Environmental Law Reporter* 20, 10337–10344.

Gardner, R.C. (2002). "Corps' New Regulatory Guidance Letter: Trick or Treat?" *National Wetlands Newsletter* 24:2, 3–4, 14.

Gardner, R.C. (2007). "*Rapanos* and Wetland Mitigation Banking," in K. Wroth, ed., *The Supreme Court and the Clean Water Act: Five Essays*. Royalton, VT: Vermont Law School's Land Use Institute.

234 Selected References and Further Reading

Houck, O.A. (1989). "Hard Choices: The Analysis of Alternatives under Section 404 of the Clean Water Act and Similar Environmental Laws." *University of Colorado Law Review* 60, 773–840.

Hough, P., and M.M. Robertson. (2009). "Mitigation under Section 404 of the Clean Water Act: Where It Comes From, What It Means." *Wetlands Ecology and Management* 17:1, 15–33.

Jacobs, A.J. (2007). *The Year of Living Biblically: One Man's Humble Quest to Follow the Bible as Literally as Possible.* New York: Simon and Schuster.

Municipality of Anchorage v. United States. (1992). 980 F.2d 1320 U.S. Court of Appeals for the Ninth Circuit.

Natural Resource Law Institute. (1988). "A Guide to Federal Wetlands Protection Under Section 404 of the Clean Water Act." *AnadromoU.S. Fish Law Memo*, Issue 46. Portland, OR: Lewis and Clark Law School.

Rodgers, W.H. (1994). *Environmental Law* 2nd ed. St. Paul, MN: West.

Stewart v. Potts (1998) 996 F.Supp. 668 U.S. District Court for the Southern District of Texas.

U.S. Army Corps of Engineers. (1990). "Permit Elevation, Old Cutler Bay Associates." www.epa.gov/owow/wetlands/pdf/CutlerBayGuidance.pdf.

U.S. Department of the Army and U.S. Environmental Protection Agency. (1990). Memorandum of Agreement Between the Department of the Army and the Environmental Protection Agency Concerning the Determination of Mitigation under the Clean Water Act Section 404(b)(1) Guidelines. Washington, D.C..

U.S. Environmental Protection Agency. (2008a). "Final Determination of the Assistant Administrator for Water Pursuant to Section 404(c) of the Clean Water Act Concerning the Proposed Yazoo Backwater Area Pumps Project in Issaquena County, MS." *Federal Register* 73:183, 54398–54400.

U.S. Environmental Protection Agency. (2008b). "Proposed Determination to Prohibit, Restrict, or Deny the Specification, or the Usefor Specification, of an Area as a Disposal Site; Yazoo River Basin, Issaquena County, MS." *Federal Register* 73:54, 14806–14820.

U.S. Environmental Protection Agency. (2010a). *Chronology of 404(c) Actions.* www.epa.gov/owow/wetlands/regs/404c.html.

U.S. Environmental Protection Agency. (2010b). *Chronology of 404(q) Actions.* www.epa.gov/wetlands/guidance/404q.html.

von Hermann, D., ed. (2006). *Resorting to Casinos: The Mississippi Gambling Industry.* Jackson, MS: University of Mississippi Press.

Weissman, S., J. Willmuth, and F. Clemente. (1999). *Betting on Trent Lott: The Casino Gambling Industry's Campaign Contributions Pay Off in Congress.* Washington, D.C.: Public Citizen's Congress Watch.

Winter, L. (1990). "Sununu Pulled Rank." *National Wetlands Newsletter* 12:2, 3–7, 8.

Zabel v. Tabb (1970) 430 F.2d 199 U.S. Court of Appeals for the Fifth Circuit.

Chapter 6

Barringer, F. (2006). "Fewer Marshes + More Man-Made Ponds = Increased Wetlands." *New York Times*, March 31.

Bernard S. (2010). "Some Common Myths About Original TABASCO® brand Pepper Sauce." www.tabasco.com/tabasco_history/faqs.cfm.

Carter-Finn, K., A.W. Hodges, D.J. Lee, and M.T. Olexa. (2006). "Benefit-Cost Analysis of Melaleuca Management in South Florida." Gainesville, FL: Food and Resource Economics Department, Florida Cooperative Extension Service, Institute of Food and Agricultural Sciences, University of Florida.

Colbert Nation. (2006). Colbert Report—The Word—Birdie. www.colbertnation .com/the-colbert-report-videos/61258/april-04-2006/the-word—birdie.

Dahl, T.E. (1990). *Wetlands Losses in the United States 1780's to 1980's*. Washington, D.C.: U.S Department of the Interior, U.S. Fish and Wildlife Service.

Dahl, T.E. (2006). *Status and Trends of Wetlands in the ConterminoU.S. United States 1998 to 2004*. Washington, D.C.: U.S. Department of the Interior, U.S. Fish and Wildlife Service.

Dahl, T.E. and C.E. Johnson. (1991). *Status and Trends of Wetlands in the ConterminoU.S. United States, Mid-1970's to Mid-1980's*. Washington, D.C.: U.S. Department of the Interior, Fish and Wildlife Service.

Dale, V.H., K.L. Kline, J. Wiens, and J. Fargione. (2010). "Biofuels: Implications for Land Useand Biodiversity." *Biofuels and SU.S.tainability Reports* Washington, D.C.: Ecological Society of America.

Dray, F.A., B.C. Bennett, and T.D. Center. (2006). "Invasion History of Melaleuca quinquenervia (Cav.) S.T. Blake in Florida." *Castanea* 71:3, 210–225.

Florida Department of Environmental Regulation. (1991). *Report of the Effectiveness of Permitted Mitigation*.

Franke, J.M. (2007). *The Invasive Species Cookbook: Conservation through Gastronomy*. Wauwatosa, WI: Bradford Street Press.

Frayer, W.E., T.J. Monahan, D.C. Bowden, and F.A. Graybill. (1983). *Status and Trends of Wetlands and Deepwater Habitats in the ConterminoU.S. United States, 1950's to 1970's*. Petersburg, FL: National Wetlands Inventory, U.S. Fish and Wildlife Service, St.

Gardner, R.C. (1996). "Banking on Entrepreneurs: Wetlands, Mitigation Banking, and Takings." *Iowa Law Review* 81:3, 527–587.

Gardner, R.C. (2005). "Mitigation," in K.D. Connolly, S.M. Johnson, and D.R. Williams, eds., *Wetlands Law and Policy: Understanding Section 404*. Chicago: American Bar Association.

Harvey, R.G., M.L. Brien, M.S. Cherkiss, M. Dorcas, M. Rochford, R.W. Snow, and F.J. Mazzotti. (2008). "Burmese Pythons in South Florida: Scientific Support for Invasive Species Management." Gainesville, FL: Wildlife Ecology and Conservation Department, Florida Cooperative Extension Service, Institute of Food and Agricultural Sciences, University of Florida.

Louisiana Department of Wildlife and Fisheries. (2010). "Recipes." http://web.wlf
.louisiana.gov/experience/nutriacontrol/humanconsumption/recipes.cfm.

Mazzotti, F.J., T.D. Center, F.A. Dray, and D. Thayer. (1997). "Ecological Consequences of Invasion by *Melaleuca Quinquenervia* in South Florida Wetlands:
Paradise Damaged, not Lost." Gainesville, FL: Wildlife Ecology and Conservation Department, Florida Cooperative Extension Service, Institute of Food
and Agricultural Sciences, University of Florida.

Mullin, B.H., L.W.J. Anderson, J.M. DiTomaso, R.E. Eplee, and K.D. Getsinger.
(2000). "Invasive Plant Species." Issue Paper No. 13. Ames, IA: Council for
Agricultural Science and Technology.

National Research Council. (2001). *Compensating for Wetland Losses under the Clean
Water Act*. Washington D.C.: National Academies Press.

Pittman, C. and M. Waite. 2005. "They won't say no." *St. Petersburg Times*, May 22.

Ruhl, J.B., S.E. Kraft, and C.L. Lant. (2007). *The Law and Policy of Ecosystem Services*.
Washington, D.C.: Island Press.

Sierra Club v. U.S. Army Corps of Engineers. (2002). 295 F.3d 1209 U.S. Court of
Appeals for the Eleventh Circuit.

Stedman, S., and T.E. Dahl. (2008). *Status and Trends of Wetlands in the Coastal Watersheds of the Eastern United States 1998 to 2004*. Washington, D.C.: National
Oceanic and Atmospheric Administration, National Marine Fisheries Service
and U.S. Department of the Interior, Fish and Wildlife Service.

Streever, B. (2001). *Saving Louisiana? The Battle for Coastal Wetlands*. Jackson, MS:
University Press of Mississippi.

U.S. Department of Agriculture, Natural Resources Conservation Service. (2010).
Final Benefit-Cost Analysis for the Wetlands Reserve Program (WRP). http://www
.nrcs.usda.gov/programs/farmbill/2008/bca-cria/WRP_BCAnalysis-FINAL-5
-17-10.pdf.

U.S. Environmental Protection Agency. (2010). *Recent Compensatory Mitigation
Evaluations and Reports*. www.epa.gov/wetlandsmitigation/#evaluations.

U.S. Environmental Protection Agency and U.S. Fish and Wildlife Service. (1994).
*Interagency Follow-Through Investigation of Compensatory Wetland Mitigation
Sites*.

Vileisis, A. (1997). *Discovery of the Unknown Landscape: A History of America's Wetlands*. Washington, D.C.: Island Press.

Wiebe, J. and E. Mouton. (2009). *Nutria Harvest and Distribution 2008–2009 and
A Survey of Nutria Herbivory Damage in Coastal Louisiana in 2009*. New Iberia,
LA: Coastal and Nongame Resources, Louisiana Department of Wildlife and
Fisheries.

Williams, D.R. (2005). "Agricultural Programs," in K.D. Connolly, S.M. Johnson,
and D.R. Williams, eds., *Wetlands Law and Policy: Understanding Section 404*.
Chicago: American Bar Association.

Zedler, J. (2007). "Success: An Unclear, Subjective Descriptor of Restoration Outcomes." *Ecological Restoration* 25:3, 162–168.

Chapter 7

Apogee Research, Inc. (1994). *An Examination of Wetlands Programs: Opportunities for Compensatory Mitigation*. Alexandria, VA: Institute for Water Resources.

Brumbaugh, R. and R.T. Reppert. (1994). *National Wetlands Mitigation Banking Study First Phase Report*. Alexandria, VA: Institute for Water Resources.

Brumbaugh, R., and F. Tabatabai. (1998). *National Wetland Mitigation Banking Study: The Early Mitigation Banks: A Follow-up Review*. Alexandria, VA: Institute for Water Resources.

Environmental Law Institute. (1993). *Wetland Mitigation Banking*. Washington, D.C.

Environmental Law Institute. (2002). *Banks and Fees: The Status of Off-site Wetland Mitigation in the United States*. Washington, D.C.

Environmental Law Institute and Institute for Water Resources. (1994). *Wetland Mitigation Banking: Resource Document*. Alexandria, VA: Institute for Water Resources.

Gardner, R.C. (1996). "Banking on Entrepreneurs: Wetlands, Mitigation Banking, and Takings." *Iowa Law Review* 81:3, 527–587.

Gardner, R.C. (2003). "Rehabilitating Nature: A Comparative Review of Legal Mechanisms That Encourage Wetland Restoration Efforts." *Catholic University Law Review* 52, 527–620.

Gardner, R.C., and T.J.P. Radwan. (2005). "What Happens When a Wetland Mitigation Bank Goes Bankrupt?" *Environmental Law Reporter* 35:4, 10590–10604.

Gatewood, S. (1995). "Disney Banks on Mitigation." *National Wetlands Newsletter* 17:5, 7–9.

Haynes, W.J., and R.C. Gardner. (1993). "The Value of Wetlands as Wetlands: The Case for Mitigation Banking." *Environmental Law Reporter* 23:5, 10261–10265.

In re IT Group, Inc. (2006). 399 Bankruptcy Reporter 338, U.S. District Court for the District of Delaware.

National Research Council. (2001). *Compensating for Wetland Losses under the Clean Water Act*. Washington, D.C.: National Academies Press.

Pittman, C., and M. Waite. (2009). *Paving Paradise: Florida's Vanishing Wetlands and the Failure of No Net Loss*. Gainesville, FL: University Press of Florida.

Reiss, K.C., E. Hernandez, and M. Brown. (2007). An Evaluation of the Effectiveness of Mitigation Banking in Florida: Ecological Success and Compliance with Permit Criteria. www.dep.state.fl.US/water/wetlands/docs/mitigation/Final_Report.pdf.

Robertson, M.M. (2006). "Emerging Ecosystem Service Markets: Trends in a Decade of Entrepreneurial Wetland Banking." *Frontiers in Ecology and the Environment* 4:6, 297–302.

Robertson, M.M. (2008). "The Entrepreneurial Wetland Banking Experience in Chicago and Minnesota." *National Wetlands Newsletter* 30:4, 14–17, 20.

Ruhl, J.B. and J. Salzman. (2006). "The Effects of Wetland Mitigation Banking on People." *National Wetlands Newsletter* vol 28:2, pp1, 9–14.

Scodari, P., L. Shabman, and D. White. (1995). *National Wetlands Mitigation Banking Study—Commercial Wetland Mitigation Credit Markets: Theory and Practice*. Alexandria, VA: Institute for Water Resources.

Shabman, L., P. Scodari, and D. King. (1994). *National Wetlands Mitigation Banking Study—Expanding Opportunities for Successful Mitigation: The Private Credit Market Alternative*. Alexandria, VA: Institute for Water Resources.

Sibbing, J.M. (2005). "Mitigation Banking: Will the Myth Ever Die?" *National Wetlands Newsletter* 27:6, 5–6, 8.

Transportation Equity Act for the 21st Century (1998) Public Law 105–178.

U.S. Department of the Army and U.S. Environmental Protection Agency. (1990). Memorandum of Agreement between the Department of the Army and the Environmental Protection Agency Concerning the Determination of Mitigation under the Clean Water Act Section 404(b)(1) Guidelines. Washington, D.C..

U.S. Department of the Army, U.S. Environmental Protection Agency, U.S. Department of Agriculture, U.S. Department of Interior, and U.S. Department of Commerce. (1995). "Federal Guidance for the Establishment, Use and Operation of Mitigation Banks." *Federal Register* vol 60:228, pp58605–58614.

U.S. Environmental Protection Agency and U.S. Department of the Army. (1993). Memorandum to the Field, Subject: Establishment and Useof Wetland Mitigation Banks in the Clean Water Act Section 404 Regulatory Program. Washington, D.C..

U.S. Federal Highway Administration, U.S. Environmental Protection Agency, and U.S. Army Corps of Engineers. (2003). Federal Guidance on the Useof the TEA-21 Preference for Mitigation Banking to Fulfill Mitigation Requirements under Section 404 of the Clean Water Act. Washington, D.C.

Wilkinson, J. and J. Thompson. (2006). *2005 Status Report on Compensatory Mitigation in the United States*. Washington, D.C.: Environmental Law Institute.

Chapter 8

Environmental Law Institute. (2002). *Banks and Fees: The Status of Off-site Wetland Mitigation in the United States*. Washington, D.C.; Environmental Law Institute

Gardner, R.C. (2000). "Money for Nothing? The Rise of Wetland Fee Mitigation." *Virginia Environmental Law Journal* 19:1, 1–56.

Gardner, R.C. (2008). "Legal Considerations," in N. Carroll, J. Fox, and R. Bayon, eds., *Conservation & Biodiversity Banking*. Earthscan, London.

National Research Council. (2001). *Compensating for Wetland Losses under the Clean Water Act*. Washington, D.C.: National Academies Press.

The Nature Conservancy. (2010). *Virginia Aquatic Resources TrU.S.t Fund Annual Report—2009*. www.deq.virginia.gov/export/sites/default/wetlands/pdf/VARTF_2009_Annual_Report.pdf.

Teresa, S. (2006). "The Demise of The Environmental TrU.S.t." Ecosystem Marketplace, www.ecosystemmarketplace.com/pages/dynamic/article.page.php?page_id=4227&section=home&eod=1.

Urban, D.T. and J.H. Ryan. (1999). "A Lieu-Lieu Policy with SerioU.S. Shortcomings." *National Wetlands Newsletter* 21:4, 5, 9.

U.S. Department of the Army, U.S. Environmental Protection Agency, U.S. Department of Agriculture, U.S. Department of Interior, and U.S. Department of Commerce. (1995). "Federal Guidance for the Establishment, Useand Operation of Mitigation Banks." *Federal Register* vol 60:228, pp58605–58614.

U.S. Department of the Army, U.S. Environmental Protection Agency, U.S. Fish and Wildlife Service, and National Oceanic and Atmospheric Administration. (2000). Federal Guidance on the Useof In-Lieu-Fee Arrangements for Compensatory Mitigation under Section 404 of the Clean Water Act and Section 10 of the Rivers and Harbors Act. Washington, D.C.

U.S. General Accounting Office. (2001). *WETLANDS PROTECTION: Assessments Needed to Determine Effectiveness of In-Lieu-Fee Mitigation*. Washington, D.C.

U.S. General Accounting Office. (2004). *Principles of Federal Appropriations Law*, 3rd ed., vol II, Washington, D.C.

Wilkinson, J. and J. Thompson. (2006). *2005 Status Report on Compensatory Mitigation in the United States*. Environmental Law Institute, Washington, D.C.

Chapter 9

Environmental Law Institute. (2010). "Wetlands Program—Conferences, Trainings, & Seminars." www.eli.org/Program_Areas/wetlands_events.cfm.

Gardner, R.C. (1996). "Banking on Entrepreneurs: Wetlands, Mitigation Banking, and Takings." *Iowa Law Review* 81:3, 527–587.

Gardner, R.C. (2002). "Corps' New Regulatory Guidance Letter on Mitigation: Trick or Treat?" *National Wetlands Newsletter* 24:2, 3–4, 14.

Gardner, R.C. (2007). "Reconsidering In-Lieu Fees: A Modest Proposal." Ecosystem Marketplace, www.ecosystemmarketplace.com/pages/dynamic/article.page.php?page_id=5073&section=home&eod=1.

Gardner, R.C., J. Zedler, A. Redmond, R.E. Turner, C.A Johnston, V.R. Alvarez, C.A. Simenstad, K.L. Prestegaard, and W.J. Mitsch. (2009). "Compensating for Wetland Losses under the Clean Water Act (Redux): Evaluating the Federal Compensatory Mitigation Regulation." *Stetson Law Review* 38:2, 213–249.

Murphy, J., J. Goldman-Carter, and J. Sibbing. (2009). "New Mitigation Rule Promises More of the Same: Why the New Corps and EPA Mitigation Rule

Will Fail to Protect Our Aquatic Resources Adequately." *Stetson Law Review* vol 38:2, pp311–336.

National Mitigation Banking Association v. U.S. Army Corps of Engineers (2007) No. 06-cv-2820, 2007 Westlaw 495245, U.S. District Court for the Northern District of Illinois.

National Research Council. (2001). *Compensating for Wetland Losses under the Clean Water Act*. Washington, D.C.: National Academies Press.

Ruhl, J.B., J. Salzman, and I. Goodman. (2009). "Implementing the New Ecosystem Services Mandate of the Section 404 Compensatory Mitigation Program—A Catalyst for Advancing Science and Policy." *Stetson Law Review* vol 38:2, pp251–272.

Strand, M. (2009). "Do the Mitigation Regulations Satisfy the Law? Wait and See." *Stetson Law Review* 38:2, 273–310.

Teresa, S. (2009). "Perpetual Stewardship Considerations for Compensatory Mitigation and Mitigation Banks." *Stetson Law Review* 38:2, 337–356.

U.S. Army Corps of Engineers. (2002). "Guidance on Compensatory Mitigation Projects for Aquatic Resource Impacts under the Corps Regulatory Program Pursuant to Section 404 of the Clean Water Act and Section 10 of the Rivers and Harbors Act of 1899." RGL 02-2, Washington, D.C.

U.S. Department of the Army and U.S. Environmental Protection Agency. (1990). Memorandum of Agreement Between The Department of the Army and The Environmental Protection Agency Concerning the Determination of Mitigation Under the Clean Water Act Section 404(b)(1) Guidelines. Washington, D.C.

U.S. Department of the Army, U.S. Environmental Protection Agency, U.S. Department of Commerce, U.S. Department of Interior, U.S. Department of Agriculture, and U.S. Department of Transportation. (2002). "National Wetlands Mitigation Action Plan." www.epa.gov/owow/wetlands/pdf /map1226withsign.pdf.

U.S. Department of Defense and Environmental Protection Agency. (2008). "Compensatory Mitigation for Losses of Aquatic Resources; Final Rule." *Federal Register* 73:70, 19594–19705.

U.S. Department of Defense and Environmental Protection Agency. (2006). "Compensatory Mitigation for Losses of Aquatic Resources; Proposed Rule." *Federal Register* 71:59, 15520–15556.

U.S. Federal Aviation Administration, U.S. Air Force, U.S. Army, U.S. Environmental Protection Agency, U.S. Fish and Wildlife Service, and U.S. Department of Agriculture. (2003). Memorandum of Agreement Between the Federal Aviation Administration, the U.S. Air Force, the U.S. Army, the U.S. Environmental Protection Agency, the U.S. Fish and Wildlife Service, and the U.S. Department of Agriculture to Address Aircraft-Wildlife Strikes. www.epa .gov/owow/wetlands/pdf/FAAmitigationmoa.pdf.

Chapter 10

Gardner, R.C. (1996). "Banking on Entrepreneurs: Wetlands, Mitigation Banking, and Takings." *Iowa Law Review* 81:3, 527–587.

Gardner, R.C. (2000). "Money for Nothing? The Rise of Wetland Fee Mitigation." *Virginia Environmental Law Journal* 19:1, 1–56.

Gwaltney v. Chesapeake Bay Foundation. (1987). 484 U.S. 49, U.S. Supreme Court.

Lucas v. United States. (2008). 516 F.3d 316 U.S. Court of Appeals for the Fifth Circuit.

Northwest Environmental Defense Center v. U.S. Army Corps of Engineers. (2000). 118 F.Supp.2d 1115 U.S. District Court for the District of Oregon.

Rapanos v. United States. (2006). 547 U.S. 715, U.S. Supreme Court.

Rodgers, W.H. (2009). *Rodgers' Environmental Law*. West, St. Paul, MN.

Strand, M.N. (2009). *Wetlands Deskbook*, Third Edition. Environmental Law Institute, Washington, D.C.

U.S. Attorney's Office, Southern District of Texas. (2009). "Civil Penalty Is Imposed Against Galveston Resident for Violating Clean Water Act." www.Justice.gov/usao/txs/releases/March%202009/030309Scruggs.htm.

U.S. Department of the Army and U.S. Environmental Protection Agency. (1989). Memorandum between the Department of the Army and The Environmental Protection Agency, Federal Enforcement for the Section 404 Program of the Clean Water Act. Washington, D.C.

U.S. Department of Justice. (2005). "Defendants Receive Major Jail Sentences, Pay Restitution for Major Wetlands Criminal Prosecution." www.justice.gov/opa/pr/2005/December/05_enrd_644%20.html.

U.S. General Accounting Office. (1988). *Wetlands The Corps of Engineers' Administration of the Section 404 Program*. Washington, D.C.

United States v. Cumberland Farms of Connecticut, Inc. (1987) 826 F.2d 1151 U.S. Court of Appeals for the First Circuit.

United States v. Cundiff. (2009). 555 F.3d 200, U.S. Court of Appeals for the Sixth Circuit.

United States v. Hawkins. (2005). Amended Complaint, Civil No. 3:05CV-12-H, U.S. District Court for the Western District of Kentucky.

United States v. Hawkins. (2006). Consent Judgment, Civil No. 3:05CV-12-H, U.S. District Court for the Western District of Kentucky.

United States v. Pozsgai. (1993). 999 F.2d 719, U.S. Court of Appeals for the Third Circuit.

United States v. Reaves. (1996). 923 F.Supp. 1530, U.S. District Court for the Middle District of Florida.

United States v. Savoy Senior HousingCorporation. (2008). Civil Action No. 6:06-cv-00031(nk) Consent Decree U.S. District Court for the Western District of Virginia.

United States v. Van Leuzen. (1993). 816 F. Supp. 1171, U.S. District Court for the Southern District of Texas.

Chapter 11

Bowles v. United States (1991) 23 Cl.Ct. 443 U.S. Court of Claims.

Cooley v. United States. (2003). 324 F.3d 1297, U.S. Court of Appeals for the Federal Circuit.

Florida Rock Industries, Inc. v. United States. (1999). 45 Fed.Cl. 21, U.S. Court of Federal Claims.

Forest Properties, Inc. v. United States. (1999). 177 F.3d 1360, U.S. Court of Appeals for the Federal Circuit.

Gardner, R.C. (1996). "Banking on Entrepreneurs: Wetlands, Mitigation Banking, and Takings." *Iowa Law Review* vol 81:3, pp527–587.

Good v. United States. (1999). 189 F.3d 1355, U.S. Court of Appeals for the Federal Circuit.

Kelo v. City of New London. (2005). 545 U.S. 469, U.S. Supreme Court.

Laguna Gatuna, Inc. v. United States. (2001). 50 Fed.Cl. 336, U.S. Court of Federal Claims.

Loveladies Harbor, Inc. v. United States . (1994). 28 F.3d 1171, U.S. Court of Appeals for the Federal Circuit.

Lucas v. South Carolina Coastal Council. (1992). 505 U.S. 1003, U.S. Supreme Court.

Meltz, R. (2000). *Wetlands Regulation and the Law of Property Rights "Takings."* Washington, D.C.: Congressional Research Service.

Palazzolo v. Rhode Island. (2001). 533 U.S. 606, U.S. Supreme Court.

Pax Christi Memorial Gardens, Inc. v. United States. (2002). 52 Fed.Cl. 318, U.S. Court of Federal Claims.

Penn Central Transportation Co. v. New York City. (1978). 438 U.S. 104, U.S. Supreme Court.

Pennsylvania Coal Co. v. Mahon. (1922). 260 U.S. 393, U.S. Supreme Court.

Sierra Club v. Van Antwerp. (2009). 2009 WL 6409272, No. 03-23427-CIV-HOEVELER U.S. District Court for the Southern District of Florida

United States v. Riverside Bayview Homes Inc.. (1985). 474 U.S. 121, U.S. Supreme Court.

Chapter 12

Barstow, D., L. Dodd, J. Glanz, S. Saul, and I. Urbina. (2010). "Regulators Failed to Address Risks in Oil Rig Fail-Safe Device." *New York Times*, June 21.

Brooke, R., G. Fogel, A. Glaser, E. Griffen, and K. Johnson. (2009). *Corn Ethanol and Wildlife*. Washington, D.C.: National Wildlife Federation.

Connolly, K.D. (2007). "Survey Says: Army Corps No Scalian Despot." *Environmental Law Reporter* vol 37:5, pp10317–10361.

Emmert-Mattox, S., S. Crooks, and J. Findsen. (2010). "Wetland Grasses and Gases: Are Tidal Wetlands Ready for the Carbon Markets?" *National Wetlands Newsletter* 32:6, 6–10.

Environmental Law Institute. (2008). *State Wetland Permitting Programs: Avoidance and Minimization Requirements*. Washington, D.C.: Environmental Law Institute.

Fox, J. (2008). "Getting Two for One: Opportunities and Challenges in Credit Stacking," in N. Carroll, J. Fox, and R. Bayon, eds., *Conservation and Biodiversity Banking*. London: Earthscan.

Gardner, R.C., J. Zedler, A. Redmond, R.E. Turner, C.A Johnston, V.R. Alvarez, C.A. Simenstad, K.L. Prestegaard, and W.J. Mitsch. (2009). "Compensating for Wetland Losses under the Clean Water Act (Redux): Evaluating the Federal Compensatory Mitigation Regulation." *Stetson Law Review* 38:2, 213–249.

Gellman, B. (2008). *Angler: The Cheney Vice Presidency*. New York: Penguin Press.

Giles, C. (2005). "The Sky's the Limit." *National Wetlands Newsletter* 27:1, 11–14.

Gold, R., B. Casselman, and G. Chazan. (2010). "Leaking Oil Well Lacked Safeguard Device." *Wall Street Journal* April 28. http://online.wsj.com/article/SB10001424052748704423504575212031417936798.html, accessed Nov. 29, 2010.

Hassett, K.A. (2006). *Ethanol's a Big Scam, and BU.S.h Has Fallen for It*. American Enterprise Institute for Public Policy Research, www.aei.org/article/23871.

Kennedy, R.F. (2010). *Sex, Lies and Oil Spills*. The Huffington Post, www.huffingtonpost.com/robert-f-kennedy-jr/sex-lies-and-oil-spills_b_564163.html.

National Research Council. (2001). *Compensating for Wetland Losses under the Clean Water Act*. National Academies Press, Washington, D.C.

Office of Inspector General, U.S. Department of Interior. (2010). *Investigative Report: Island Operating Company et al*. Washington, D.C.

Palmer, M.A., E.S. Bernhardt, W.H. Schlesinger, K.N. Eshleman, E. Foufoula-Georgiou, M.S. Hendryx, A.D. Lemly, G.E. Likens, O.L. Loucks, M.E. Power, P.S. White, and P.R. Wilcock. (2010). "Mountaintop Mining Consequences." *Science* vol 327, pp148–149.

Program Evaluation Division, North Carolina General Assembly. (2009). *Department of Environment and Natural Resources Wetland Mitigation Credit Determinations*. Raleigh, NC.

Urbina, I. (2010). "U.S. Said to Allow Drilling Without Needed Permits." *New York Times*, May 14, pA1.

UN-REDD Programme. (2010). *The United Nations Collaborative Programme on Reducing Emissions from Deforestation and Forest Degradation in Developing Countries*. www.un-redd.org/.

Wilkinson, J.B., J.M. McElfish, R. Kihslinger, R. Bendick, and B.A. McKenney. (2009). *The Next Generation of Mitigation: Linking Current and Future Mitigation Programs with State Wildlife Action Plans and Other State and Regional Plans*. Washington, D.C.: Environmental Law Institute and The Nature Conservancy.

Epilogue

Goldstein, M.R. and M.J. Goldstein. (1993). "Inverse Condemnation." *New York Law Journal* 210:121, 3.

Reiss, K.C., E. Hernandez, and M. Brown. (2007). An Evaluation of the Effectiveness of Mitigation Banking in Florida: Ecological Success and Compliance with Permit Criteria. www.dep.state.fl.US/water/wetlands/docs/mitigation/Final_Report.pdf.

INDEX

Island Press | Board of Directors